Fundamentals of Analytical Chemistry

Fundamentals of Analytical Chemistry

Olivia Price

MURPHY & MOORE
www.murphy-moorepublishing.com

Published by Murphy & Moore Publishing
1 Rockefeller Plaza,
New York City, NY 10020, USA
www.murphy-moorepublishing.com

Fundamentals of Analytical Chemistry
Olivia Price

International Standard Book Number: 978-1-63987-242-8 (Hardback)

Cataloging-in-Publication Data

Fundamentals of analytical chemistry / Olivia Price.
p. cm.
Includes bibliographical references and index.
ISBN 978-1-63987-242-8
1. Analytical chemistry. 2. Chemistry. I. Price, Olivia.
QD75.22 .F86 2022
543--dc23

Contents

Preface

The branch of chemistry which deals with the analysis of substances is called analytical chemistry. It involves the use of classical methods such as Kastle-Meyer test, flame tests, and gravimetric and volumetric analysis. Modern methods in this field include spectroscopy, calorimetry and electrochemical methods which are used for separation, identification and quantification of matter. Analytical chemistry is further divided into two broad areas, namely, quantitative analysis and qualitative analysis. Quantitative analysis determines the absolute or relative quantity regarding the concentration of one or more substances present in a sample or compound. Qualitative analysis focuses on determining the quality of a particular compound, irrespective of its quantity or concentration. The principles of analytical chemistry have widespread use in the food and beverage industry, chemical industry, pharmaceutical industry and agricultural industry. The topics included in this book on analytical chemistry are of utmost significance and bound to provide incredible insights to readers. While understanding the long-term perspectives of the topics, it makes an effort in highlighting their impact as a modern tool for the growth of the discipline. Those in search of information to further their knowledge will be greatly assisted by this book.

A detailed account of the significant topics covered in this book is provided below:

Chapter 1- The branch of chemistry which deals with the study of separation, identification and quantification of matter is termed as analytical chemistry. It includes quantitative analysis, qualitative analysis, titrimetric analysis, gravimetric analysis and chemical equilibrium. This is an introductory chapter which will introduce briefly all the significant aspects of analytical chemistry.

Chapter 2- Chromatography includes the techniques used to separate different components from a mixture. There are various types of chromatography such as gas chromatography, liquid chromatography, affinity chromatography, ion exchange chromatography, paper chromatography, etc. This chapter discusses in detail these various types of chromatography.

Chapter 3- Spectroscopy refers to the study of absorption and emission of radiations by matter. The various types of spectroscopy are mass spectrometry, molecular spectroscopy, atomic absorption spectroscopy, atomic emission spectroscopy, flame emission spectroscopy and Raman spectroscopy. This chapter has been carefully written to provide an easy understanding of these types of spectroscopy.

Chapter 4- Thermal analysis is a technique used to analyze the materials on the basis of their change with temperature. Some of the thermoanalytical methods include differential thermal analysis, differential scanning calorimetry and thermogravimetric methods. The chapter closely examines these methods of thermal analysis to provide an extensive understanding of the subject.

Chapter 5- The methods which are used to study an analyte by calculating the potential and current of an electrochemical cell holding the analyte are known as electroanalytical methods. Science and technology have undergone rapid developments in the past decade which has resulted in the discovery of significant electroanalytical methods; which have been extensively detailed in this chapter.

Chapter 6- There are various techniques which are used for separation such as chromatography, extraction, filtration, decantation, sublimation, evaporation, simple distillation, fractional distillation, centrifugation and magnetic separation. The diverse applications of these chemical separation techniques have been thoroughly discussed in this chapter.

It gives me an immense pleasure to thank our entire team for their efforts. Finally in the end, I would like to thank my family and colleagues who have been a great source of inspiration and support.

Olivia Price

1

Analytical Chemistry: An Introduction

The branch of chemistry which deals with the study of separation, identification and quantification of matter is termed as analytical chemistry. It includes quantitative analysis, qualitative analysis, titrimetric analysis, gravimetric analysis and chemical equilibrium. This is an introductory chapter which will introduce briefly all the significant aspects of analytical chemistry.

Analytical chemistry is as old as chemistry. Analytical chemistry acts as a foundation for other branches of chemistry. In fact the science of chemistry came into existence as a result of human inquisitiveness to understand the nature of an extraordinary variety of matter that surrounded him, and this knowledge was obtained by analysing different types of material. A complex material on analysis can give rise to a number of new and simpler constituents. Each of the constituents can be further analysed and if this process is continued, a stage will come when a constituent will not give anything new; such a constituent is called an element. Thus the concept of an element as a substance which cannot be broken down into something simpler by ordinary chemical methods arose from analytical data on different substances.

The construction of chemical balance provided a quantitative aspect to chemical analysis. With this development the study of analytical chemistry stimulated quantitative approach to various problems of chemistry. Chemical reactions were studied on the basis of qualitative and also quantitative changes that occurred; this resulted in the discovery of several new compounds and the five laws of chemical combination. In an attempt to theoretically explain these laws, Dalton developed the atomic theory which played a tremendous role in developing chemistry into a quantitative experience and was mainly preparative in nature centred on empirical synthetic methods. The induction of analytical approach brought about a revolutionary transformation from magic and alchemy to quantitative scientific chemistry.

Steps in Analytical Procedure

Structural information about a complex compound can be obtained either by preparing it from some simple constituents or by breaking it chemically into smaller and simpler units and then identifying them. The former approach involves synthesis of the compound while the latter is termed chemical analysis.

- Inorganic and Organic Analysis: Chemical analysis is broadly classified as inorganic or organic depending upon the nature of the material under examination. Elemental analysis deals with the detection and determination of various elements present in a compound. Functional group analysis involves the determination of certain groupings of atoms such as carboxyl group (-COOH) or hydroxyl group (-OH) in an organic material.
- Major, Minute and Trace Constituents: A major constituent is one whose amount is 1 per cent or more of the sample material. A minor constituent present in quantities smaller than 0.01 percent is called a trace constituent.

- Complete and Partial Analysis: Chemical analysis is said to be complete when it involves the determination of all the components detected qualitatively in the sample. The analysis is partial when it aims at determining only one or a few of the components of the sample such as determination of copper in a copper ore.
- Classical Chemical Analysis: Quantitative chemical analysis started with the application of techniques of gravimetric procedures. In a gravimetric measurement, a known volume of sample solution is treated with an excess of a suitable reagent which quantitatively precipitates the desired constituent present in the sample solution. The precipitate which is of known concentration is filtered, washed, dried and weighed. Knowing the weight of the precipitate, the amount of the desired constituent in the test solution is calculated. For example, an excess of dilute sulphuric acid is added to a given solution containing barium ions. The precipitated of barium sulphate formed is filtered, washed, dried and weighed. From the weight of barium sulphate the quantity of barium in the given solution is calculated. Because such determinations are based on the measurement of weight, these are referred to as gravimetric determinations.

In electro-gravimetric analysis, the constituent to be determined is deposited on the electrode by passing electric current through a suitable electrolytic cell. For example, the amount of copper in given copper sulphate solution can be determined by passing current through the solution and weighing the copper deposited at the negative electrode. Here, electric current acts as a precipitating agent.

Gravimetric determination of a volatile constituent can be done by heating the sample and recording the loss of weight. Gravimetric procedures are quite accurate but are lengthy and tedious. Another group of techniques was soon developed in which quantitative analysis was achieved by measuring volume of solutions; hence it was called volumetric analysis. In this type of analysis, to the sample solution of unknown concentration, a reagent solution of known concentration is gradually added till the reaction between them is just complete as shown by some indicator.

The volume of the sample and reagent solutions are known, the concentration of the reagent solution is also known so the concentration of the given sample solution can be calculated. For example, a known volume of hydrochloric acid solution whose concentration is to be determined is taken in a conical flask and a few drops of phenolphthalein solution are added as indicator. A solution of sodium hydroxide of known concentration is gradually added through a burette until the solution in the flask becomes just pink. The volume of sodium hydroxide solution added is recorded and from this, the concentration of given hydrochloric acid is calculated. This process is called titration and the determination is termed titrimetric determination.

Certain substances, under suitable experimental conditions, quantitatively liberate a gas. The measurement of the volume of the liberated gas can be used as a basis for determining the substances producing the gas. Such analytical procedures are called “gasometric methods”. For example, semicarbazide can be decomposed with lead peroxide and the liberated nitrogen can be measured. Knowing the volume of nitrogen, the amount of semicarbazide present in the sample solution can be calculated.

A particular component of a gaseous mixture can be absorbed in a suitable absorbent. The decrease in the volume of the gaseous mixture gives the volume of constituent absorbed. This method of analysis is called "gas analysis". For example, a mixture of oxygen and carbon dioxide gases can be passed through a solution of potassium hydroxide when carbon dioxide alone is absorbed. Thus the decrease in the volume of the gaseous mixture will be equal to the volume of carbon dioxide present in the mixture.

Gravimetric methods along with titrimetric procedures belong to the classical methods of analysis. Titrimetric procedures are much simpler and convenient and hence a large number of titration methods have been worked out for determining a wide variety of inorganic and organic substances. In direct titrimetric methods, sample solution is directly titrated with reagent solution. The indirect procedure consists of adding a known excess of reagent and titrating back the unused reagent.

Evaluation of Analytical Chemistry

A remarkable fact about the evaluation of analytical chemistry is that workers in other branches such as inorganic, organic, biochemistry etc., have also contributed significantly to its growth. The difference in approach is that whereas an analytical chemist have his main interest in the methods and techniques themselves, other workers develop analytical procedures for their own specific problems. For example, a worker in the field of chemical kinetics is not primarily interested in various analytical methods and techniques but may develop a method for quantitatively analysing a substance whose concentration is to be determined in order to obtain the necessary kinetic data. On the other hand main interest of an analytical chemist is in the development and improvement of an analytical procedure itself and in testing its reliability. He also studies the interference caused by the presence of other substances and attempt to modify the procedure so that such interference can be eliminated. It is necessary to distinguish between an analytical chemist and an analyst. An analytical chemist carefully chooses a chemical reaction and uses it for developing an analytical procedure taking into account various theoretical considerations. He also studies the effect of different factors that can influence the result of the determination. The job of an analyst is simply to follow the given instructions to perform a determination.

The earlier methods of quantitative analysis were those involving gravimetric procedures. Soon thereafter the technique of volumetric analysis emerged which due to its inherent simplicity and rapidity received preference over gravimetric methods which were lengthy and tedious. One of the earliest volumetric determinations was developed by Marguerite who titrated ferrous iron with potassium permanganate. Such a procedure is now more appropriately called a titrimetric procedure rather than a volumetric method because the latter is a more general term including gasometric methods and gas analysis which are also based on the measurement of volume. Due to their inherent simplicity titrimetric analyses have found and continue to find extensive applications. Both direct and indirect titrimetric procedures have been worked out for the quantitative analysis of wide variety of organic as well as inorganic compounds with the modern accent on micro and submicro analysis.

Over the last 50 years or so there has been a growing tendency to make use of certain instruments to achieve quantitative analysis. For example, instrumental techniques such as potentiometric, conductometric, photometric, amperometric etc., have been applied to locate the end point in a titration or to follow the course of a chemical reaction. It should be noted that in such cases the titrimetric methods are not basically altered from their standard procedures, the instrument simply act as a substitute for an indicator. Recently such methods commonly known as the instrumental methods of analysis have been increasingly used especially in the field of industrial and commercial quantitative analysis due to their rapidity and sensitivity. But it would be a mistake to conclude that the so called chemical methods would be totally eliminated from considerations.

The classical and physicochemical methods of quantitative analysis should be regarded as complementary. Because once an instrument to be used in quantitative analysis has been properly calibrated, it can be used with great advantage to achieve rapid analysis and due to high sensitivity of physicochemical procedures; samples at microgram level can be handled.

The advantages of microchemical methods are saving of time, labour and material. Much of the work on vitamins, hormones and other natural products could be done due to the development of microanalytical methods because many of the compounds were present in micro quantities. In about last 50 years there has been increasing sophistication in all areas of chemistry, physics and biological sciences. This created analytical problems which required use of sophisticated instrumentation for their solution. For example, in determining traces of impurities at part per billion level or determining traces of pollutants in the atmosphere of industrial area.

Analytical Problems and their Solutions

The solutions of all analytical problems, whether qualitative or quantitative, follow the same basic pattern. This may be described under seven general headings:

- Selection: The selection of the method of analysis is an important step in the solution of an analytical problem. A choice cannot be made until the overall problem is defined, and where possible a decision should be taken by the client and the analyst in consultation. Inevitably, in the method selected, a compromise has to be reached between the sensitivity, precision and accuracy desired of the results and the costs involved. For example, X-ray, fluorescence, spectrometry may provide rapid but rather imprecise quantitative results in a trace element problem. Atomic absorption spectrophotometry, on the other hand, will supply more precise data, but at the expense of more time-consuming chemical manipulations.
- Sampling: Correct sampling is the cornerstone of reliable analysis. The analyst must decide in conjunction with technological colleagues how, where, and when a sample should be taken so as to be truly representative of the parameter that is to be measured.
- Preliminary sample treatment: For quantitative analysis, the amount of sample taken is usually measured by mass or volume. When a homogeneous sample already exists, it may be subdivided without further treatment. With many solids such as ores, however, crushing and mixing are prior requirements. The sample often needs additional preparation for analysis, such as drying, ignition and dissolution.
- Separation: A large proportion of analytical measurements are subject to interference from other constituents of the sample. Newer methods increasingly employ instrumental

techniques to distinguish between analyte and interference signals. However, such distinction is not always possible and sometimes a selective chemical reaction can be used to mask the interference.

- Final measurement: This step is often the quickest and easiest of the seven but can only be as reliable as the preceding stages. The fundamental necessity is a known proportionality between the magnitude of the measurement and the amount of analyte present.
- Method validation: It is pointless carrying out the analysis unless the results obtained are known to be meaningful. This can only be ensured by proper validation of the method before use and subsequent monitoring of its performance. The analysis of validated standards is the most satisfactory approach. Validated standards have been extensively analysed by a variety of methods, and an accepted value for the analyte obtained. A standard should be selected with a matrix similar to that of the sample. In order to ensure continued accurate analysis, standards must be reanalysed at regular intervals.
- The assessment of results: Results obtained from an analysis must be assessed by the appropriate statistical methods and their meaning considered in the light of the original problem.

Importance of Analytical Chemistry to Various Branches of Science

- Analytical chemistry finds wide applications not only in different branches of chemistry but also in other physical and biological sciences and in many fields of engineering.
- Geologists use analytical procedures for analysing ground water, minerals, rocks, ores etc. In agriculture, chemical analysis is used to determine the composition of soils, in the production of fertilizers, insecticides and weed killers. Medical and biological research programmes depend on chemical analysis which provides useful information that enhances our understanding of vital processes and helps in developing medicines to cure various diseases.
- In order to safeguard public health there is constant checking of foods, drugs, cosmetics, water supplies etc., and this is done in analytical laboratories. Waste disposals and the composition of air in industrial areas are analysed to know the extent of harm they would cause to public health so that necessary preventive steps can be taken.
- There is hardly a branch of national economy which does not make use of analytical techniques. Chemical analysis is important in controlling the quality of raw materials, intermediate and finished products. Hence to produce high quality products it is essential to have analytical control at all the stages of technological processes. The sale of raw materials by suppliers and their purchase by users is by analysis metallurgical products are most essential materials of modern economy. The properties of alloys depend on its composition which is established by analytical methods.
- A very large number of examples can be cited where analytical chemistry finds application. It would be no exaggeration. If it is said that there is hardly a material related to modern living in which analytical chemistry did not play some part. With the development of sophisticated instruments, analytical techniques can now be used to determine traces of impurities at the part per billion levels.

Quantitative Analysis

The quantitative determination of a substance by precipitation, followed by isolation and weighing of the precipitate is called gravimetric analysis. Quantitative analysis is used to determine the percentage of a particular element or ion in a sample. In a typical gravimetric analysis the percentage of an ion of interest in a solid compound is determined. This procedure involves dissolving a substance of unknown composition in water and allowing the ion of interest to react with a counter ion to form a precipitate. The precipitate is then isolated, dried and weighed.

Knowing the mass and the chemical formula of the precipitate the mass of the ion of interest can be determined from the per cent mass composition of the precipitate. From the masses of the ion of interest and the original compound, the percentage of the ion of interest in the original compound can be calculated.

Quantitative Analysis of Organic Compounds

The estimation of percentage composition of the various elements present in a compound is the next step in determining the formula of the compound. The different methods employed for the estimation of various elements are described below:

Estimation of Carbon and Hydrogen

These two elements are always estimated together by Liebig's combustion method. A weighted amount of the compound is heated strongly with excess copper oxide in an atmosphere of air or oxygen. The constituent's hydrogen and carbon are thus oxidised to water and carbon dioxide, which are collected separately and weighed. The percentage of carbon and hydrogen in the compound can be calculated as given below:

% C = 12/44 x Mass of CO_2 formed/Mass of the substance x 100

% H = 1/18 x Mass of H_2 formed/Mass of the substance x 100

Estimation of Nitrogen

Dumas Method

This method can be applied to all organic compounds containing nitrogen. The method is based on the principle that if an organic compound containing nitrogen is heated with copper oxide, free nitrogen and oxides of nitrogen are formed along with other products (carbon dioxide, water vapour etc). The oxides of nitrogen are reduced to free nitrogen on passing over heated copper and the whole of nitrogen is collected over KOH solution. The volume of nitrogen collected is measured and from this the percentage present in the compound is calculated.

% N = 28/22400 x Volume of nitrogen at NTP/Mass of compound x 100

Kjeldahl's Method

This method is commonly employed for estimating nitrogen in food materials and fertilizers.

Though this method is simpler than Dumas method, it is not applicable to all nitrogenous organic compounds.

Estimation of Halogen

Carius Method

A known mass of the organic substance containing halogen is heated with fuming nitric acid along with few crystals of silver nitrate in a sealed tube. The silver halide is formed which o=is separated, washed dried and weighed. From the mass of silver halide obtained the percentage of halogen is calculated.

% Halogen = Atomic mass of halogen x Mass of silver halide obtained x 100/Molar mass of silver halide x Mass of compound taken.

Carius method does not give satisfactory results with iodine as silver iodide is slightly soluble in nitric acid and some iodine is also produced even in the presence of excess silver nitrate. Moreover, for highly halogenated aromatic compounds, the results are not accurate.

Estimation of Sulphur

Sulphur is also estimated by carius method. In this case, the organic compound is heated only with nitric acid. Sulphur present in the compound is thus oxidized to sulphuric acid which is treated with barium chloride to precipitate barium sulphate. The precipitate of barium sulphate is washed, dried and weighed. From the weighted mass of barium sulphate obtained, the percentage of sulphur is calculated.

% S = 32/133 x Mass of $BaSO_4$/Mass of the compound x 100

Estimation of Phosphorus

Phosphorus is estimated like sulphur, an organic compound is heated with fuming nitric acid. Phosphorus in the compound is thus oxidized to phosphoric acid which is precipitated by adding magnesia mixture. The precipitate of magnesium ammonium phosphate is ignited to obtain magnesium pyrophosphate.

$2MgNH_4PO_4$ (Magnesium ammonium phosphate) $\rightarrow Mg_2P_2O_7 + 2NH_3 + H_2O$

Magnesium pyrophosphate is weighed and the percentage of phosphorus calculation is,

%P = 62/222 x Mass of $Mg_2P_2O_7$/Mass of the compound x 100

Determination of Molecular Formula of the Compound

The molecular formula of a compound expresses the actual number of atoms of various elements present in a molecule of the compound. Molecular formula may be either the same as the empirical formula or a simple multiple of it.

Molecular formula = (Empirical formula) x n

Where n is a whole number. The value of n is obtained by dividing the molecular mass by empirical formula mass.

n = Molecular mass/Empirical formula mass.

Classification of Methods of Quantitative Analysis

The methods of quantitative analysis can be classified from different points of view based on the nature of material under examination, the type of method employed, the amount of desired constituent in sample material and so on.

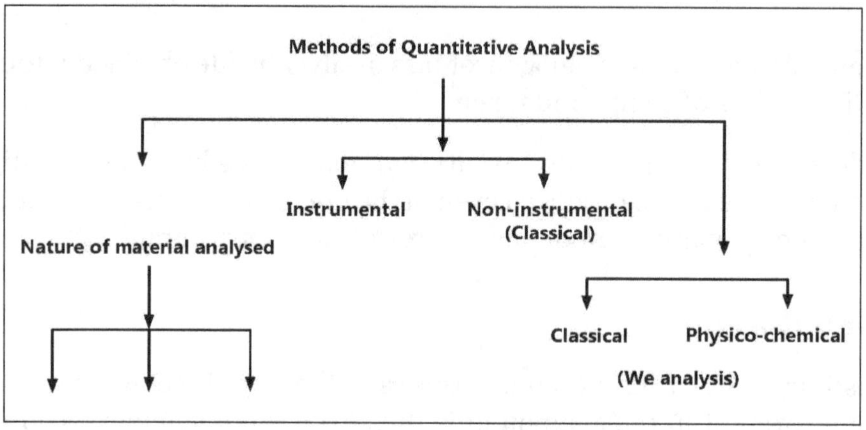

- The analysis can be termed organic, inorganic, biochemical etc., depending on the nature of the material analysed.
- Another classification of quantitative analysis is non-instrumental and instrumental. The former includes gravimetric and titrimetric procedures whereas the latter involves use of instruments such as colorimeter, a conductivity meter or a potentiometer and so on.
- Another classification into classical and physico-chemical methods of analysis is also not very sound theoretically although it is quite useful for the sake of convenience. According to this classification, gravimetric and titrimetric methods are termed chemical methods though they are based on the measurement of physical properties, viz. weight and volume. The methods of analysis based on such physical properties as potential, conductance, current strength, optical rotation, etc., are generally referred to as physio-chemical methods.
- The classical chemical analysis which consists of gravimetric and titrimetric method is also known as wet analysis. There are certain other procedures which are based on matter-energy interaction such as, a colorimetric determination which involves passage of light a form of energy through solution of the substance, to be determined, which is matter.
- The methods of quantitative analysis can also be classified on the basis of the size of the sample for a determination. The term macro-analysis is used when the determination involves 0.1 g or more of the sample. If the amount of the sample is approximately 0.01 to 0.1 g, the method is called semi-micro and for samples weighing 0.001-0.01 g, the term micro-method is used. Ultra-micro analysis involves samples containing less than 0.001g of material; some authors have used the term sub-micro analysis also. Certain procedures

have been described for analysing quantities smaller than those handled in ultra-micro analysis; these constitute what is known as nano analysis.

- According to another classification, macro-methods are those in which sample contains more than 2 milliequivalents (meq) of the material and 0.1 N solutions are employed in the determination. In semi-micro methods, the sample contains about 1 meq of substance and 0.05 to 0.1 N solutions are used in its titrimetric determination. When the amount of the substance is around 0.1 meq the method used is called a micro-method; this generally involves the use of 0.01 N reagent solutions.

Qualitative Analysis

Qualitative analysis is the determination of the chemical composition of a sample. It encompasses a set of analytical chemistry techniques that provide non-numerical information about a specimen. Qualitative analysis can tell you whether an atom, ion, functional group, or compound is present or absent in a sample, but it doesn't provide information about its quantity. Quantification of a sample, in contrast, is called quantitative analysis.

The flame test is a technique used to identify ions using qualitative analysis.

Techniques and Tests

Qualitative analysis involves chemical tests, such as the Kastle-Meyer test for blood or the iodine test for starch. Another common qualitative test, used in inorganic chemical analysis, is the flame test. Qualitative analysis typically measures changes in color, melting point, odor, reactivity, radioactivity, boiling point, bubble production, and precipitation. Methods include distillation, extraction, precipitation, chromatography, and spectroscopy.

Branches of Qualitative Analysis

The two main branches of qualitative analysis are organic qualitative analysis (such as the iodine test) and inorganic qualitative analysis (such as the flame test). Inorganic analysis looks at the elemental and ionic composition of a sample, usually by examination of ions in aqueous solution.

Organic analysis tends to look at types of molecules, functional groups, and chemical bonds. Example: She used qualitative analysis to find that the solution contained Cu^{2+} and Cl^{-} ions.

Gravimetric Analysis

Gravimetric analysis is a method in analytical chemistry to determine the quantity of analyte based on the mass of a solid. Example: Measuring the solids suspended in the water sample – Once a known volume of water is filtered, the collected solids are weighed.

The Principle of Gravimetric Analysis

The principle behind the gravimetric analysis is that the mass of an ion in a pure compound and can be determined. Later, used to find the mass percent of the same ion in a known quantity of an impure compound.

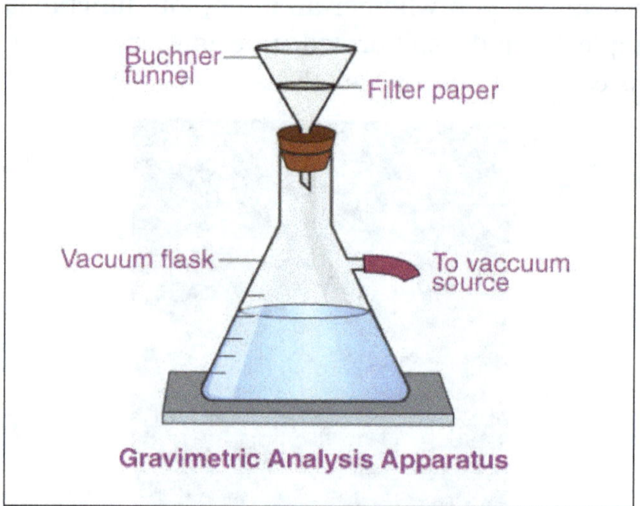

Gravimetric Analysis Apparatus.

Steps followed in the Gravimetric Analysis:

- Preparation of a solution containing a known weight of the sample.
- Separation of the desired constituent.
- Weighing the isolated constituent.
- Computation of the amount of the particular constituent in the sample from the observed weight of the isolated substance.

Types of Gravimetric Analysis

There are four fundamental types of gravimetric analysis. Of which, there are two common types involving changes in the phase of the analyte to separate it from the rest of a mixture, resulting in a change in mass.

- Volatilization gravimetry: Volatilization Gravimetry involves separating components of our mixture by heating or chemically decomposing the sample.
- Precipitation gravimetry: Precipitation Gravimetry uses a precipitation reaction to separate one or more parts of a solution by incorporating it into a solid.
- Electrogravimetry: Electrogravimetry is a method used to separate and quantify ions of a substance, usually a metal.
- Thermogravimetric: Thermogravimetric is a method of thermal analysis in which changes in physical and chemical properties of materials are measured as a function of increasing temperature or as a function of time.

Advantages of Gravimetric Analysis

If the methods are followed carefully, it provides exceedingly precise analysis. It is used to determine the atomic masses of many elements to six-figure accuracy. It provides a little room for instrumental error and does not require a series of standards for calculation of an unknown.

Disadvantages of Gravimetric Analysis

It usually provides only for the analysis of a single element, or a limited group of elements, at a time. Comparing modern dynamic flash combustion coupled with gas chromatography with traditional combustion analysis.

Chemical Equilibrium

Chemical equilibrium is a condition in the course of a reversible chemical reaction in which no net change in the amounts of reactants and products occurs. A reversible chemical reaction is one in which the products, as soon as they are formed, react to produce the original reactants. At equilibrium, the two opposing reactions go on at equal rates, or velocities, hence there is no net change in the amounts of substances involved. At this point the reaction may be considered to be completed; i.e., for some specified reaction condition, the maximum conversion of reactants to products has been attained.

The conditions that pertain to equilibrium may be given quantitative formulation. For example, for the reversible reaction $A \leftrightharpoons B + C$, the velocity of the reaction to the right, r_1, is given by the mathematical expression (based on the law of mass action) $r_1 = k_1(A)$, where k_1 is the reaction-rate constant and the symbol in parentheses represents the concentration of *A*. The velocity of the reaction to the left, r_2, is $r_2 = k_2(B)(C)$. At equilibrium, $r_1 = r_2$, therefore:

$$k_1(A)_e = k_2(B)_e(C)_e \quad \text{or}$$
$$k_1 / k_2 = (B)_e(C)_e / (A)_e = K.$$

The subscript e represents conditions at equilibrium. For a given reaction, at some specified condition of temperature and pressure, the ratio of the amounts of products and reactants present at equilibrium, each raised to their respective powers, is a constant, designated the equilibrium constant of the reaction and represented by the symbol K. The value of the equilibrium constant varies with the temperature and pressure according to the principle of Le Chatelier.

By methods of statistical mechanics and chemical thermodynamics, it can be shown that the equilibrium constant is related to the change in the thermodynamic quantity called the standard Gibbs free energy accompanying the reaction. The standard Gibbs free energy of the reaction, $\Delta G°$, which is the difference between the sum of the standard free energies of the products and that of the reactants, is equal to the negative natural logarithm of the equilibrium constant multiplied by the so-called gas constant R and the absolute temperature T:

$$\Delta G° = -RT \ln K$$

The equation allows the calculation of the equilibrium constant, or the relative amounts of products and reactants present at equilibrium, from measured or derived values of standard free energies of substances.

References

- Quantitative-analysis, chemistry: byjus.com, Retrieved 11, July 2020
- Definition-of-qualitative-analysis-604626: thoughtco.com, Retrieved 02, February 2020
- Gravimetric-analysis: byjus.com, Retrieved 23, August 2020
- Chemical-equilibrium, science: britannica.com, Retrieved 09, March 2020

2

Chromatography: Theory and Methods

Chromatography includes the techniques used to separate different components from a mixture. There are various types of chromatography such as gas chromatography, liquid chromatography, affinity chromatography, ion exchange chromatography, paper chromatography, etc. This chapter discusses in detail these various types of chromatography.

Chromatography is a method for separating the constituents of a solution (gas or liquid) by exploiting the different properties of different molecules. The technique employs a mobile phase (gas or liquid) to transport the solution to be analyzed through the stationary phase (solid or liquid) which absorbs or impedes different components of the solution to different degrees and thus causes their separation as different layers. It is an invaluable tool in the hands of analytical scientists for the separation and quantification of components in a mixture of organic compounds. Since its evolution chromatography has found wide use in separation of components of mixtures ranging from simplest gases to most complex hydrocarbon mixtures containing hundreds of different compounds. Samples can be gaseous, liquids or even solids which are readily soluble in suitable solvents.

Due to its versatility the applications cover complete range of chemical compounds having diverse characteristics such as boiling range, molecular weights, volatilities, and thermal stabilities. Gas chromatography, Liquid chromatography, Thin-layer chromatography, Super critical fluid chromatography and hyphenated techniques such as GC – MS, and LC – MS cover a vast number of applications in diverse areas such as pharmaceuticals, material development, foods, petroleum products and forensic investigations requiring very high resolution and detection limits. A few application of chromatography in different industries are:

Chemical Industry

- Contaminants' detection in pesticides and oils.
- Water sample testing.
- Air quality checks.
- Numerous applications in life sciences.

Pharmaceutical Industry

- Taking element composition and its molecular weight to separate compounds.
- During the process of drug development.
- Detecting chemicals or trace elements in various samples with proper analysis.
- Examination of mixture purity.
- Identification of unknown compounds.

Forensic Industry

- Blood and hair sample analysis under crime scene testing.
- Forensic pathology.

Food Industry

- Identifying the food products' nutritional values and quality.
- Detection of additives and food spoilage.

Molecular Biology Studies

- The fuel industry, biochemical processes, and biotechnology use HPLC for purification and fractionation process.
- Metabolomics and proteomics study through specific chromatography techniques.
- Nucleic acid research with specific chromatography methods.

These are only the popular applications of chromatography. The technique is used in various other places and industries.

Basic Working of Chromatography

The exact process of chromatography will depend on where you want to use it and which method you opt for. However, we can still define a basic procedure that consists of simple four steps:

- Step 1: Careful exposure of a specific analyte amount to the mobile phase stream that was already running.
- Step 2: Then, the analyte is carried out through the stationary phase with the help of the mobile phase.
- Step 3: Finally, the separation takes place when the analyte components react with the stationary phase at different levels. Some of them react more, while others don't.
- Step 4: The separated out analyte components are then again carried out by the mobile phase. It then takes the components to a separate instrument. There, they get quantified with the proper detection of their presence.

And in these four steps, the process gets complete. We can say that chromatography is divided into three basic components: Mobile Phase, Stationary Phase, and Separated Molecules.

Chromatography Parts

Chromatography Columns

Separation of sample components before detection is the essence of chromatographic techniques. A chromatographic system uses a column to achieve the desired separation. A column comprises of a tube packed with a stationary phase on which separation takes place based on physico – chemical interactions of separating compounds with the stationary phase. The mobile phase or the carrier gas elutes less weakly retained components first followed by more strongly retained components. The column dimensions and composition is based on the chromatographic technique selected and the degree of required separation. In general HPLC columns are shorter and wider than GC columns.

Chromatographic Detectors

After separation the individual components reach the detector which provides response in terms of the area or peak height depending on the amount of the eluting compound. Each chromatographic separation technique offers a range of detectors depending on the nature of eluting compounds. Detection can be specific for a particular compound or a range of compounds or it can depend on some physical properties such as reflective index of the mobile phase. Such detectors are referred to as bulk property detectors in comparison to selective or specific detectors.

Chromatographic Data

Chromatographic separations appear as peaks in the chromatogram except for thin-layer chromatography for which separate zones are seen on the plate. Chromatographic peaks are separated in time in the chromatogram depending on the nature of separating compounds and the separation efficiency of the column. Sharp well resolved peaks indicate high degree of column resolution whereas broad or overlapping peaks are indicative of poor resolution. Each peak represents a sample component and is characterized by its retention time under defined operating conditions. For purposes of quantification the peak height or more accurately the area under the peak defines the amount of the component present in the mixture.

Quantitation methods require percentage composition of a particular component in the mixture and such estimations are made on the presumption that the detector responds to all the components equally and peaks are generated for each compound.

$$\% \; of \text{ analyte } = \frac{\text{Area of analyte peak}}{\text{sum total area of all peaks}} \times 100$$

Quite often the amount of analyte is specified in relation to another compound which is referred to as a reference standard having established traceability. Such compounds should be available in pure form and have chemical identity which is same or close to the identity of the analyte to be determined. Modern-day analytical systems come equipped with sophisticated application software which is capable of operational control, data interpolation and calculations to achieve the desired results.

Theory of Chromatography

Let us consider first a small section of a column to be used for liquid-liquid partition chromatography. The partitioning of sample molecules between the two phases is based on the distribution coefficient, K (or the partition coefficient) and can be explained by equation below:

$$K = C_s / C_M$$

where the subscripts S and M pertain to the stationary and mobile phases, respectively.

The distribution coefficient, K, defined in above equation applies to many kinds of mechanisms. Nevertheless, whether the process is ion exchange, solution, adsorption, or vaporization from solution, the value of K determines the relative populations in the two phases. If the value of K is large, the population in the stationary phase is larger than that in the mobile phase and the molecules of that component will spend more time, on the average, in the stationary phase.

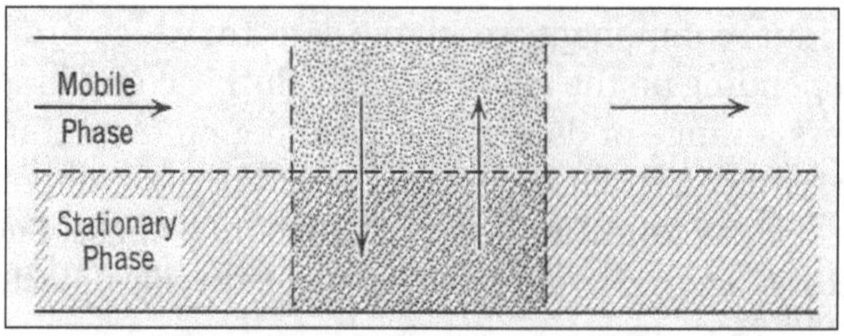

Section of a chromatographic column: The stationary phase is located in the lower half of the column for clarity.

The fraction of the total time that an average molecule spends in the mobile phase is directly related to the fraction of the total population which is found in the mobile phase.

$$\text{Fraction of time spent in mobile phase} = \frac{\text{No. of molecules in mobile phase}}{\text{Total no. of molecules}}$$

$$= \frac{C_M V_M}{C_M V_M + C_S V_S}$$

$$= \frac{1}{1 + K V_S / V_M}$$

$$= \frac{1}{1 + k'}$$

where $k' = KV_S / V_M$, a new variable called the capacity factor. k' is also equal to $C_S V_S / C_M V_M$.

Rate of Travel

Equation above gives the fraction of time spent in the mobile phase, and on multiplication by the linear velocity, u, of the mobile phase we have the rate of travel of an average sample molecule:

$$\text{Rate} = u \frac{1}{1 + k'}$$

Thus, the rate of travel of an average molecule is determined by:

- Carrier mobile phase velocity: same for all components.
- Distribution coefficient: unique for each component.

Retention Time

The time it takes a component to travel the length of the column, L, is known as the retention time, t_R. From this definition and above equation we obtain,

$$t_R = \frac{\text{length}}{\text{rate}} = \frac{L}{u}(1 + k') = t_M (1 + k')$$

where t_m is the time required for a carrier molecule to traverse the column. Above equation is one of the basic retention equations for all forms of chromatography.

Retention Volume

It is often more convenient to measure the volume of mobile phase required to elute the sample component. If the volume flow rate of mobile phase, F, is constant, then the retention volume, V_R is obtained.

Volume = time × flow rate

$$V_R = V_m(1+k') = V_m + KV_s$$

Substituting above equation into equation $t_R = \frac{\text{length}}{\text{rate}} = \frac{L}{u}(1+k') = t_M(1+k')$ we obtain,

$$V_R = V_M(1+k') = V_M + KV_s$$

where V_M is the volume of mobile phase contained within the column at a given instant. Alternatively, V_M is the void or empty space in the column not occupied by the stationary phase. Above equation is another form of the basic retention equation and is applicable to all types of chromatography. If the stationary phase is a solid, V_S may be converted to its area (adsorption) or its ion exchange capacity, or any other appropriate unit.

Relative Retention

We can eliminate the effects of the operational variables if we compare the retention time of a component to that of a known compound run on the same column under the same conditions. Before taking the ratio of the two retention times, it is important to subtract the time required to elute a non-retarded component. This step is essential because all components must travel the length of the column and therefore spend the same amount of time in the mobile phase. We should compare only the times they spend in (or on) the stationary phase. The same argument applies to relative retention volume, etc. The relative retention is often given the symbol "α".

$$\alpha = \frac{t_R - t_M}{t_R^* - t_M} = \frac{V_R - V_M}{V_R^* - V_M} = \frac{K}{K^*}$$

where the asterisks refer to the standard substance.

Ideal Chromatography

In ideal chromatography, the component zones would have no tendency to spread and would elute from the outlet of the column in the same volume or time interval as they were introduced at the inlet. Since the zone of introduction can be very sharp, an ideal chromatogram would appear as a series of very sharp, completely separated peaks, even for a sample containing many components as shown in Figure.

Non-Ideal Chromatography

The real world is, never quite ideal, and one of the aspects of non-ideality is that in a collection of molecules there are random variations in behavior. The net result of non-ideal behavior is that zones which are compact and sharp at the inlet become diffuse and broad as they pass through the column, as shown in figure.

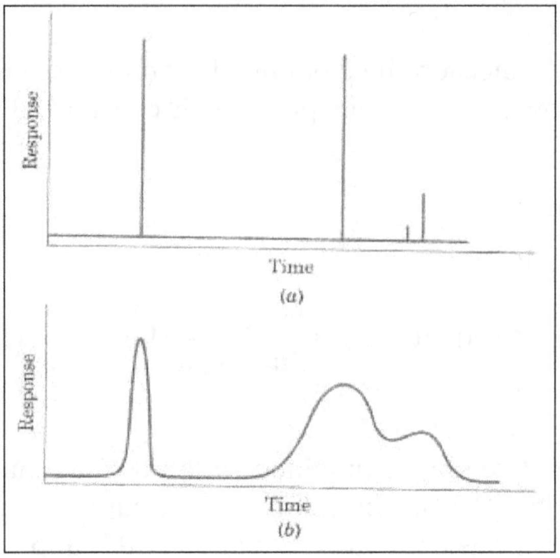

Band broadening: (a) ideal behavior; (b) non-ideal behavior.

The phenomenon of band spreading has important consequences for chromatography. Closely spaced zones which could be separated ideally do not separate completely because the width of the zones exceeds the space between them. The first step is to design a physical model, and, as usual, this ends in a compromise between a simple, easily handled model and one that attempts to take all factors into account.

Plate Theory

Martin and Synge recognized the similarity between the processes occurring in a chromatographic column and a fractional distillation column and adapted the "theoretical plate concept". However, the analogy to a Craig machine is even more direct. We can imagine that a chromatographic column consists of a large number of identical segments, or theoretical plates, in each of which an equilibration is achieved. In the Craig process, we can predict the distribution of a component after a given number of transfers (equilibrations). Conversely, in chromatography, we observe the distribution of a component along the column, or, alternatively, measure its concentration as a function of time as it reaches the column outlet and passes through a detector. To sum it up, a chromatographic column contains the same number of theoretical plates as the number of vessels in a Craig machine giving the same separation behavior. In chromatographic terms, the number of theoretical plates, n, in a column is determined from the chromatogram.

$$n = \left(\frac{4t_R}{W}\right)^2$$

Equation above requires that the peak width be measured at the intersections of the tangents with the base line. Since n is dimensionless, both t_R and W must be measured in the same units.

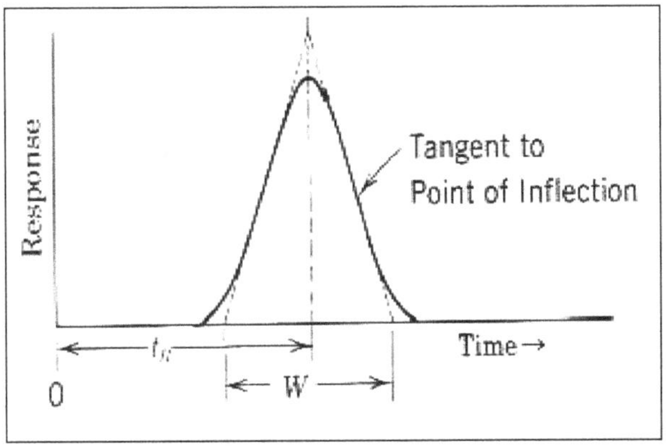

Schematic chromatogram.

The number of theoretical plates in a column, computed by above equation is not solely a function of how the column is constructed, but also of the nature of the solute, flow rate, temperature, method of sample introduction, and other variables. For a given retention time, the larger the n value, the narrower the peaks, and the closer the approach to ideal behavior. There are two ways of increasing the number of theoretical plates in a column. First, since each theoretical plate occupies the same space in the column, the number of plates is directly proportional to the column length, L. Therefore, if we double L, we double n. This also must double t_R, according to equation $t_R = \frac{\text{length}}{\text{rate}} = \frac{L}{u}(1+k') = t_M(1+k')$. Thus, all retention times (and the separations between peaks) will be doubled.

In other words, the actual gain in separation is roughly a square root function of the column length. A better approach to ideal behavior is to increase the number of equilibrations within the same time span. In other words, we should increase the number of theoretical plates per unit column length.

Rate Theory

The behavior of a collection of molecules in a chromatographic column is indeed chaotic; however, we recognize three important deviations from ideality which cause H to be greater than its ideal value of zero.

- As the molecules move from inlet to outlet of the column, they take random paths of different lengths. This is true whether the column is packed or empty.
- Molecules tend to diffuse from regions of high concentration to low.
- The continuous flow of the mobile phase causes a departure from equilibrium: the ratio C_S/C_M will always be smaller than K at the leading edge of a zone and larger than K at the trailing edge. This time lag also causes the zone to spread in both directions. Each of these three factors operates more or less independently, and their total effect on H is a summation of three terms:

$$H = \underset{\text{(Eddy diffusion)}}{\text{Contribution from non-equal paths}} + \underset{\text{(Longitudinal diffusion)}}{\text{Contribution from diffusion along column}} + \underset{\text{(Mass transfer)}}{\text{Contribution from nonequilibrium}}$$

A group of Dutch chemists derived an expression known as the Van Deemter equation:

$$H = A + B/u + Cu$$

Each of the three terms on the right side of above equation is more complex than shown here. A, Band C can be regarded as constants for a given column.

Resolution

In practice, the most important objective is to be able to separate the sample components. One is often concerned with the degree of separation, or resolution, of compounds of similar character. The two components in figure are poorly resolved and overlap to an extent that makes it impossible to identify or determine the amount of either.

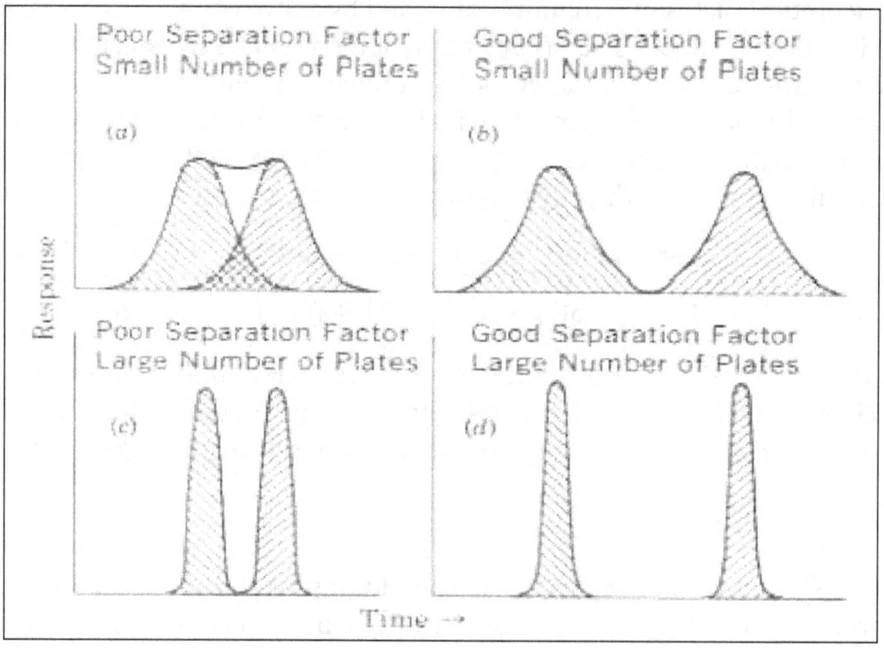

Effect of separation factor and number of theoretical plates in the column on the resolution of two peaks.

To improve the resolution, we can either (a) change the temperature or the nature of the phases to give a greater separation between the peaks as in Figure, or (b) reduce the width of the peaks by improving the column efficiency (decrease H, increase n, or (c) a combination of both (a) and (b) as in Figure, where the resolution is now more than satisfactory. Quantitatively, the resolution, R, of two peaks having retention time t. and t_2 and peak widths W_1 and W_2 is:

$$R = \frac{2(t_2 - t_1)}{W_1 + W_2}$$

The resolution of the two peaks shown in figure is about 1.05. A resolution of R > 1.5 gives so-called baseline resolution or essentially complete separation of two peaks of equal size. A resolution of R = 1.0, giving a peak overlap of approximately 2%, is usually considered to be adequate for analytical purposes.

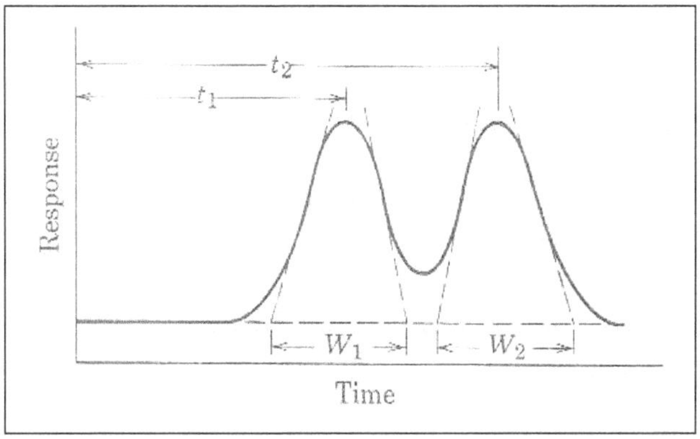

Measurements used in calculating the resolution of two peaks.

Adequate resolution may be difficult to obtain for complex samples containing many components of similar properties or for trace components. The best procedures for improving resolution vary with the type of chromatography; however, some useful generalities are in order.

To improve resolution:

Increase $t_R = t_2 - t_1$

- Increase column length, L.
- Increase the amount of stationary phase, V_s or V_L.
- Employ a better separation factor, a = g/t;
    - Decrease temperature, T.
    - Select a different stationary phase.
    - Select a different mobile phase (if liquid).

Decrease the Band Width, W

- Employ more uniform packing:
    - Pack more carefully.
    - Use smaller particles.
- Increase interfacial area between phases.
- Optimize flow rate.
- Reduce sample size.
- Reduce dead space in system.
- Reduce time constant of detector.
- Decrease diameter of column.

In developing or improving a chromatographic procedure, one or more of the following objectives may be important:

- High resolution: For complex mixtures.
- Fast analysis time: For kinetic studies or process control.
- Large throughput: For preparatory work or concentrating trace components.

All three of these objectives cannot be achieved simultaneously; for example, large samples and/or high flow rates give poor resolution. Inevitably, the operator must decide on the best compromise to meet the requirements. In making the best choices, chromatographic theory can be a great help.

Gas Chromatography

Gas chromatography differs from other forms of chromatography in that the mobile phase is a gas and the components are separated as vapors. It is thus used to separate and detect small molecular weight compounds in the gas phase. The sample is either a gas or a liquid that is vaporized in the injection port. The mobile phase for gas chromatography is a carrier gas, typically helium because of its low molecular weight and being chemically inert. The pressure is applied and the mobile phase moves the analyte through the column. The separation is accomplished using a column coated with a stationary phase.

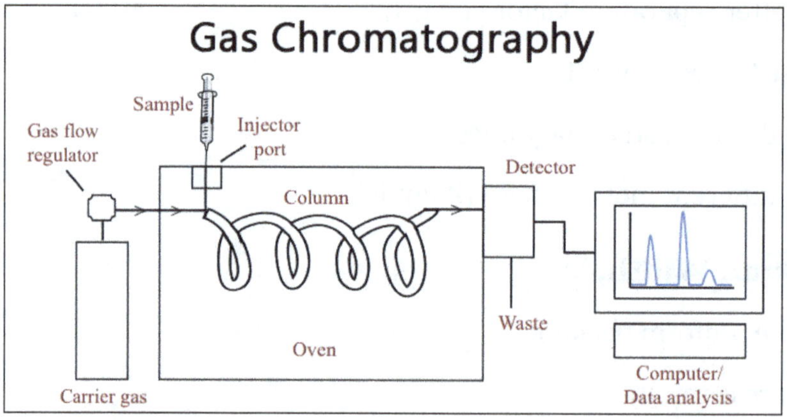

Principle of Gas Chromatography (How does Gas Chromatography Work)

The equilibrium for gas chromatography is partitioning, and the components of the sample will partition (i.e. distribute) between the two phases: the stationary phase and the mobile phase. Compounds that have a greater affinity for the stationary phase spend more time in the column and thus elute later and have a longer retention time (Rt) than samples that have a higher affinity for the mobile phase.

Affinity for the stationary phase is driven mainly by intermolecular interactions and the polarity of the stationary phase can be chosen to maximize interactions and thus the separation. Ideal peaks are Gaussian distributions and symmetrical, because of the random nature of the analyte interactions with the column.

The separation is hence accomplished by partitioning the sample between the gas and a thin layer of a nonvolatile liquid held on a solid support. A sample containing the solutes is injected into a heated block where it is immediately vaporized and swept as a plug of vapor by the carrier gas stream into the column inlet. The solutes are adsorbed by the stationary phase and then desorbed by a fresh carrier gas. The process is repeated in each plate as the sample is moved toward the outlet. Each solute will travel at its own rate through the column. Their bands will separate into distinct zones depending on the partition coefficients, and band spreading. The solutes are eluted one after another in the increasing order of their kd, and enter into a detector attached to the exit end of the column. Here they register a series of signals resulting from concentration changes and rates of elution on the recorder as a plot of time versus the composition of carrier gas stream. The appearance time, height, width, and area of these peaks can be measured to yield quantitative data.

Parts of Gas Chromatography

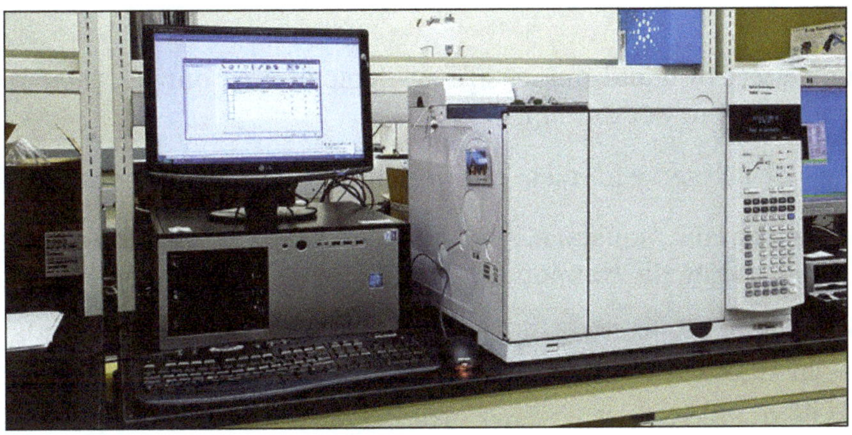

Carrier Gas in a High-Pressure Cylinder with Attendant Pressure Regulators and Flow Meters

- Helium, N_2, H, Argon are used as carrier gases.
- Helium is preferred for thermal conductivity detectors because of its high thermal conductivity relative to that of most organic vapors.
- N_2 is preferable when a large consumption of carrier gas is employed.
- Carrier gas from the tank passes through a toggle valve, a flow meter, (1-1000 ml/min), capillary restrictors, and a pressure gauge (1-4 atm).
- Flow rate is adjusted by means of a needle valve mounted on the base of the flow meter and controlled by capillary restrictors.
- The operating efficiency of the gas chromatograph is directly dependant on the maintenance of constant gas flow.

Sample Injection System

- Liquid samples are injected by a microsyringe with a needle inserted through a self-scaling, silicon-rubber septum into a heated metal block by a resistance heater.

- Gaseous samples are injected by a gas-tight syringe or through a by-pass loop and valves.
- Typical sample volumes range from 0.1 to 0.2 ml.

The Separation Column

- The heart of the gas chromatography is the column which is made of metals bent in U shape or coiled into an open spiral or a flat pancake shape.
- Copper is useful up to 250°.
- Swege lock fittings make column insertion easy.
- Several sizes of columns are used depending upon the requirements.

Liquid Phases

- An infinite variety of liquid phases are available limited only by their volatility, thermal stability and ability to wet the support.
- No single phase will serve for all separation problems at all temperatures.
    - Non-Polar: Parafin, squalane, silicone greases, apiezon L, silicone gum rubber. These materials separate the components in order of their boiling points.
    - Intermediate Polarity: These materials contain a polar or polarizable group on a long non-polar skeleton which can dissolve both polar and non-polar solutes. For example, diethyl hexyl phthalate is used for the separation of high boiling alcohols.
    - Polar - Carbowaxes: Liquid phases with a large proportion of polar groups. Separation of polar and non-polar substances.
    - Hydrogen bonding: Polar liquid phases with high hydrogen bonding e.g. Glycol.
    - Specific purpose phases: Relying on a chemical reaction with solute to achieve separations. e.g $AgNO_3$ in glycol separates unsaturated hydrocarbons.

Supports

- The structure and surface characteristics of the support materials are important parameters, which determine the efficiency of the support and the degree of separation respectively.
- The support should be inert but capable of immobilizing a large volume of liquid phase as a thin film over its surface.
- The surface area should be large to ensure the rapid attainment of equilibrium between stationary and mobile phases.
- Support should be strong enough to resist breakdown in handling and be capable of packed into a uniform bed.

- Diatomaceous earth, kieselguhr treated with Na_2CO_3 for 900 °C causes the particle fusion into coarser aggregates.
- Glass beads with a low surface area and low porosity can be used to coat up to 3% stationary phases.
- Porous polymer beads differing in the degree of cross-linking of styrene with alkyl-vinyl benzene are also used which are stable up to 250°.

Detector

- Detectors sense the arrival of the separated components and provide a signal.
- These are either concentration-dependent or mass dependant.
- The detector should be close to the column exit and the correct temperature to prevent decomposition.

Recorder

- The recorder should be generally 10 mv (full scale) fitted with a fast response pen (1 sec or less). The recorder should be connected with a series of good quality resistances connected across the input to attenuate the large signals.
- An integrator may be a good addition.

The Procedure of Gas Chromatography

Step 1: Sample Injection and Vapourization

- A small amount of liquid sample to be analyzed is drawn up into a syringe.
- The syringe needle is positioned in the hot injection port of the gas chromatograph and the sample is injected quickly.
- The injection of the sample is considered to be a "point" in time, that is, it is assumed that the entire sample enters the gas chromatograph at the same time, so the sample must be injected quickly.
- The temperature is set to be higher than the boiling points of the components of the mixture so that the components will vaporize.
- The vaporized components then mix with the inert gas mobile phase to be carried to the gas chromatography column to be separated.

Step 2: Separation in the Column

- Components in the mixture are separated based on their abilities to adsorb on or bind to, the stationary phase.
- A component that adsorbs most strongly to the stationary phase will spend the most time in the column (will be retained in the column for the longest time) and will, therefore, have the longest retention time (Rt). It will emerge from the gas chromatograph last.

- A component that adsorbs the least strongly to the stationary phase will spend the least time in the column (will be retained in the column for the shortest time) and will, therefore, have the shortest retention time (Rt). It will emerge from the gas chromatograph first.
- If we consider a 2 component mixture in which component A is more polar than component B then:
    - Component A will have a longer retention time in a polar column than component B.
    - Component A will have a shorter retention time in a non-polar column than component B.

Step 3: Detecting and Recording Results

- The components of the mixture reach the detector at different times due to differences in the time they are retained in the column.
- The component that is retained the shortest time in the column is detected first. The component that is retained the longest time in the column is detected last.
- The detector sends a signal to the chart recorder which results in a peak on the chart paper. The component that is detected first is recorded first. The component that is detected last is recorded last.

Applications

- GC analysis is used to calculate the content of a chemical product, for example in assuring the quality of products in the chemical industry; or measuring toxic substances in soil, air or water.
- Gas chromatography is used in the analysis of:
    - Air-borne pollutants.
    - Performance-enhancing drugs in athlete's urine samples.
    - Oil spills.
    - Essential oils in perfume preparation.
- GC is very accurate if used properly and can measure picomoles of a substance in a 1 ml liquid sample, or parts-per-billion concentrations in gaseous samples.
- Gas Chromatography is used extensively in forensic science. Disciplines as diverse as solid drug dose (pre-consumption form) identification and quantification, arson investigation, paint chip analysis, and toxicology cases, employ GC to identify and quantify various biological specimens and crime-scene evidence.

Advantages

- The use of longer columns and higher velocity of carrier gas permits the fast separation in a matter of a few minutes.
- Higher working temperatures up to 500 °C and the possibility of converting any material into a volatile component make gas chromatography one of the most versatile techniques.

- GC is popular for environmental monitoring and industrial applications because it is very reliable and can be run nearly continuously.
- GC is typically used in applications where small, volatile molecules are detected and with non-aqueous solutions.
- GC is favored for non-polar molecules.

Limitations

- Compound to be analyzed should be stable under GC operation conditions.
- They should have a vapor pressure significantly greater than zero.
- Typically, the compounds analyzed are less than 1,000 Da, because it is difficult to vaporize larger compounds.
- The samples are also required to be salt-free; they should not contain ions.
- Very minute amounts of a substance can be measured, but it is often required that the sample must be measured in comparison to a sample containing the pure, suspected substance known as a reference standard.

Liquid Chromatography

Liquid chromatography is particularly useful for samples containing large molecules or ionic substances with low vapor pressures, and for thermally unstable substances which cannot be vaporized without decomposing. Liquid eluents are seldom inert; therefore, the distribution coefficients depend on the chemical nature of both the stationary and the mobile phase. Liquids have much higher viscosities and greater resistance to flow than do gases. In addition rates of diffusion of solutes are five orders of magnitude slower in liquids than in gases; thus, the C term in the Van Deemter equation is much larger than the A and B terms at normal flow rates. Van Deemter plots for gas and liquid chromatography are compared in Figure. Liquid chromatography is a time consuming procedure, because low flow rates are indicated for best efficiency.

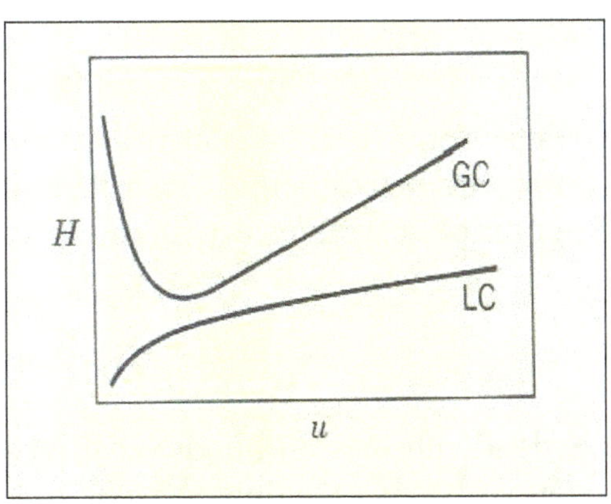

Typical Van Deemter plots for gas and liquid chromatography.

Liquid - Liquid Partition Chromatography

Partition chromatography offers great advantages over adsorption chromatography. It is far more reproducible and predictable from solubility data. Even more important, the distribution coefficient is constant over a much greater range of concentration, yielding sharper, more symmetrically shaped bands. It is the method of choice whenever a suitable solvent pair can be found.

Some care must be given to the choice of solvent pairs. They must, of course, be immiscible. The more polar of the two is normally coated on a solid support. Optimally, the fraction of a solute which is dissolved in the mobile phase should be in the range from 0.05 to 0.5; otherwise the retention time will be too long or too short. Most often, a thin film of water supported on silica gel is used as the stationary phase. Up to 50% water can be adsorbed by silica gel before it becomes too moist to handle easily. For some purposes, it may be desirable to include a buffer system in the aqueous phase to change and/or control the solubility of the solutes.

Reverse-Phase Chromatography

There are some situations in partition chromatography in which it is advantageous to use the nonpolar organic solvent as the stationary phase in order to get a more favorable value for the partition coefficient. Obviously, a nonpolar solid support is required. Powdered rubber coated with benzene is a very satisfactory stationary phase with an aqueous mobile phase.

High Performance Liquid Chromatography

In liquid chromatography, maximum efficiency is achieved at low flow rates. This is due low rates of diffusion which are characteristic of liquid phases. The alternative is to reduce the distance through which the molecules must diffuse.

Packing Materials

Closely sized particles as small as 10 μm are now becoming available. A solid glass bead of 30 to 50 μm in diameter can be coated with a layer (thin skin of 1 to 2 f-tm thickness) of porous material. These coated beads are called pellicular beads. The porous layer may serve as a solid stationary phase or be coated with a very thin layer of liquid stationary phase with an extremely large surface area.

High Pressures

Columns packed with small particles require high inlet pressures in order to give a reasonable flow rate. Pressures up to 10,000 psi are not difficult to handle in the small columns used (2 to 3 mm diam.). Improved, pulse-free pumping systems are incorporated in modern liquid chromatographs.

Detectors

Flow-through detectors with low dead volumes and high sensitivity are a necessity. Two types of detectors are currently popular. One is based on a flow-through micro-cell placed in an ultraviolet spectrophotometer. A low pressure mercury lamp is commonly used as a source and the absorption measured with the 254-nm line where most organic compounds having double bonds or aromatic groups cause at least some absorption. Measurements are not restricted to 254 nm if one employs a standard spectrophotometer. Since most common solvents do not absorb radiation in the ultraviolet and visible region, these detectors are extremely sensitive and free from interferences.

A second popular detector, the differential refractometer, continuously monitors the difference in refractive index between the pure mobile phase (reference stream) and the column effluent. These devices respond to essentially all sample components, but the refractive index differences are minute and fluctuate with small changes in temperature or composition of the mobile phase (gradient elution is difficult).

Another detector utilizes the flame ionization detector developed for gas chromatography. A continuous wire or fine chain moves through the stream emerging from the column outlet where it becomes coated with a film of the effluent. The wire then passes through an oven to evaporate the mobile phase, and finally through an oxidation oven in which the components are burned in a stream of oxygen at about 850 °C. The gaseous products are passed over a nickel catalyst bed at 350 °C which converts the carbon dioxide to methane, a substance that is readily measured by the flame ionization detector.

Many variations have been tried in an effort to overcome the difficulties of the transport system; however, these devices offer the unique advantage of responding directly to the number of carbon atoms in the component molecule. A judicious combination of these improvements has led to separations of closely related compounds in very short times, such as the impressive separation of five hydroxylated aromatics in less than 60 sec described in Figure.

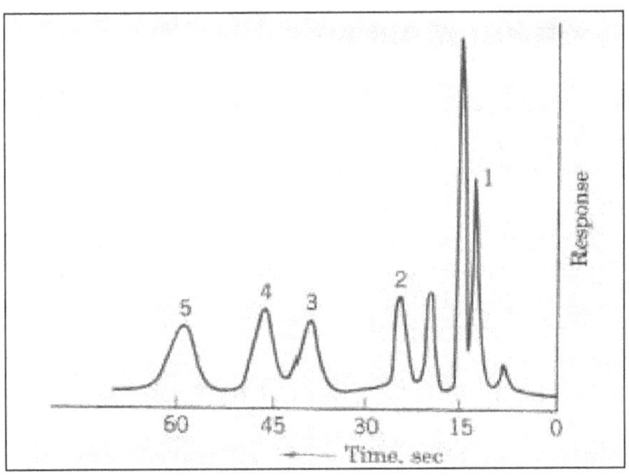

High speed separation of hydroxylated aromatics by liquid-solid chromatography.

Affinity Chromatography

Affinity chromatography is a type of liquid chromatography for the separation, purification or specific analysis of sample components. It utilizes the reversible biological interaction or molecular recognition called affinity which refers to the attracting forced exerted in different degrees between atoms which cause them to remain in combination. Example: Enzyme with and inhibitor, antigen with an antibody etc. It was discovered by Pedro Cuatrecasas and Meir Wilcheck.

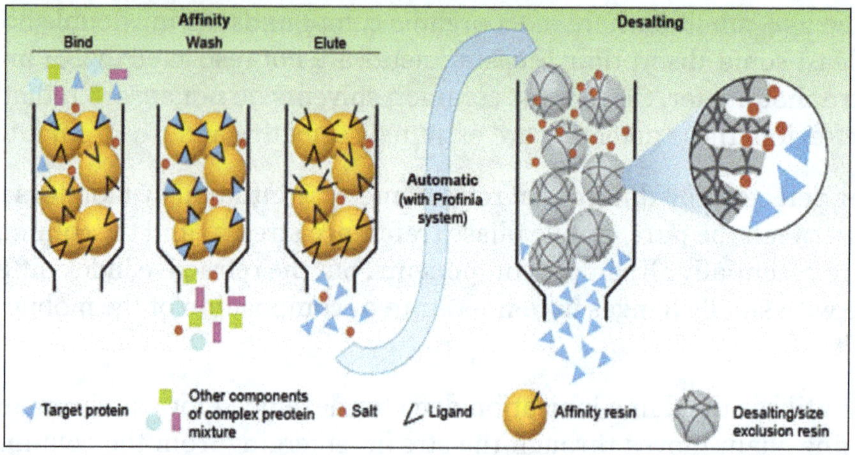

Principle of Affinity Chromatography

The stationary phase consists of a support medium, on which the substrate (ligand) is bound covalently, in such a way that the reactive groups that are essential for binding of the target molecule are exposed.

As the crude mixture of the substances is passed through the chromatography column, substances with binding site for the immobilized substrate bind to the stationary phase, while all other substances is eluted in the void volume of the column. Once the other substances are eluted, the bound target molecules can be eluted by methods such as including a competing ligand in the mobile phase or changing the pH, ionic strength or polarity conditions.

Components of Affinity Chromatography

1. Matrix: The matrix is an inert support to which a ligand can be directly or indirectly coupled. In order to for the matrix to be effective it must have certain characters:

- Matrix should be chemically and physically inert.
- It must be insoluble in solvents and buffers employed in the process.
- It must be chemically and mechanically stable.
- It must be easily coupled to a ligand or spacer arm onto which the ligand can be attached.
- It must exhibit good flow properties and have a relatively large surface area for attachment.
- The most useful matrix materials are agarose and polyacrylamide.

2. Spacer arm: It is used to improve binding between ligand and target molecule by overcoming any effects of steric hindrance.

3. Ligand: It refers to the molecule that binds reversibly to a specific target molecule. The ligand can be selected only after the nature of the macromolecule to be isolated is known. When a hormone receptor protein is to be purified by affinity chromatography, the hormone itself is an ideal candidate for the ligand. For antibody isolation, an antigen or hapten may be used as ligand. If an enzyme is to be purified, a substrate analog, inhibitor, cofactor, or effector may be used as a immobilized ligand.

Steps in Affinity Chromatography

Affinity medium is equilibrated in binding buffer. Sample is applied under conditions that favor specific binding of the target molecule(s) to a complementary binding substance (the ligand). Target substances bind specifically, but reversibly, to the ligand and unbound material washes through the column. Elution is performed specifically, using a competitive ligand, or non-specifically, by changing the pH, ionic strength or polarity. Target protein is collected in a purified, concentrated form. Affinity medium is re-equilibrated with binding buffer. These events can be summarized into the following three major steps:

- Preparation of Column:
    - The column is loaded with solid support such as sepharose, agarose, cellulose etc.
    - Ligand is selected according to the desired isolate.
    - Spacer arm is attached between the ligand and solid support.
- Loading of Sample: Solution containing a mixture of substances is poured into the elution column and allowed to run at a controlled rate.
- Elution of Ligand-Molecule Complex: Target substance is recovered by changing conditions to favor elution of the bound molecules.

Applications of Affinity Chromatography

Affinity chromatography is one of the most useful methods for the separation and purification of specific products. It is essentially a sample purification technique, used primarily for biological molecules such as proteins. Its major application includes:

- Separation of mixture of compounds.
- Removal of impurities or in purification process.
- In enzyme assays.
- Detection of substrates.
- Investigation of binding sites of enzymes.
- In in vitro antigen-antibody reactions.
- Detection of Single Nuceotide polymorphisms and mutations in nucleic acids.

Advantages of Affinity Chromatography

- High specificity.
- Target molecules can be obtained in a highly pure state.
- Single step purification.
- The matrix can be reused rapidly.

- The matrix is a solid, can be easily washed and dried.
- Give purified product with high yield.
- Affinity chromatography can also be used to remove specific contaminants, such as proteases.

Limitations of Affinity Chromatography

- Time consuming method.
- More amounts of solvents are required which may be expensive.
- Intense labour.
- Non-specific adsorption cannot be totally eliminated, it can only be minimized.
- Limited availability and high cost of immobilized ligands.
- Proteins get denatured if required pH is not adjusted.

Ion Exchange Chromatography

Ion-exchange chromatography (IEC) is part of ion chromatography which is an important analytical technique for the separation and determination of ionic compounds, together with ion-partition/interaction and ion-exclusion chromatography. Ion chromatography separation is based on ionic (or electrostatic) interactions between ionic and polar analytes, ions present in the eluent and ionic functional groups fixed to the chromatographic support. Two distinct mechanisms as follows; ion exchange due to competitive ionic binding (attraction) and ion exclusion due to repulsion between similarly charged analyte ions and the ions fixed on the chromatographic support, play a role in the separation in ion chromatography. Ion exchange has been the predominant form of ion chromatography to date. This chromatography is one of the most important adsorption techniques used in the separation of peptides, proteins, nucleic acids and related biopolymers which are charged molecules in different molecular sizes and molecular nature. The separation is based on the formation of ionic bonds between the charged groups of biomolecules and an ion-exchange gel/support carrying the opposite charge. Biomolecules display different degrees of interaction with charged chromatography media due to their varying charge properties.

Ion Exchange Mechanism

Ion-exchange chromatography which is designed specifically for the separation of differently charged or ionizable compounds comprises from mobile and stationary phases similar to other forms of column based liquid chromatography techniques. Mobil phases consist an aqueous buffer system into which the mixture to be resolved. The stationary phase usually made from inert organic matrix chemically derivative with ionizable functional groups (fixed ions) which carry displaceable oppositely charged ion. Ions which exist in a state of equilibrium between the mobile phase and stationary phases giving rise to two possible formats, anion and cation exchange are referred to as counter ion. Exchangeable matrix counter ions may include protons (H^+), hydroxide groups (OH^-), single charged mono atomic ions (Na^+, K^+, Cl^-), double charged mono atomic ions (Ca^{2+}, Mg^{2+}), and polyatomic inorganic ions (SO_4^{2-}, PO_4^{3-}) as well as organic bases (NR_2H^+) and acids (COO^-). Cations are separated on cation-exchange resin column and anions on an anion exchange resin column. Separation based on the binding of analytes to positively or negatively charged groups which are fixed on a stationary phase and which are in equilibrium with free counter ions in the mobile phase according to differences in their net surface charge.

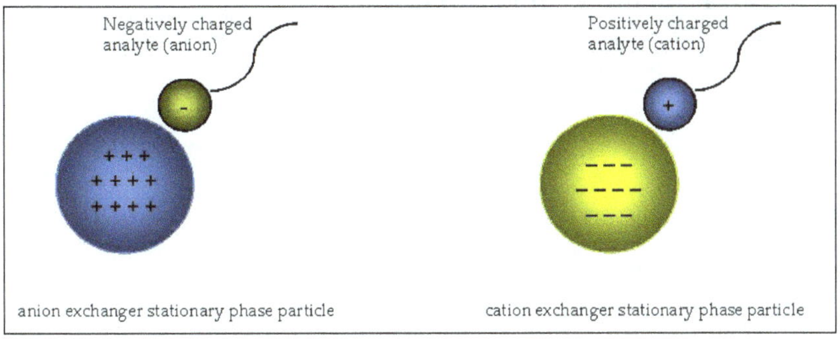

Types of ion exchangers.

Ion exchange chromatography involves separation of ionic and polar analytes using chromatographic supports derivatized with ionic functional groups that have charges opposite that of the analyte ions. The analyte ions and similarly charged ions of the eluent compete to bind to the oppositely charged ionic functional group on the surface of the stationary phase. Assuming that the exchanging ions (analytes and ions in the mobile phase) are cations, the competition can be explained using the following equation:

$$S\text{-}X^-C^+ + M^+ \leftrightarrow S\text{-}X^-M^+ + C^+$$

In this process the cation M^+ of the eluent replaced with the analyte cation C^+ bound to the anion X^- which is fixed on the surface of the chromatographic support (S).

In anion exchange chromatography, the exchanging ions are anions and the equation is represented as follow:

$$S\text{-}X^+A^- + B^- \leftrightarrow S\text{-}X^+B^- + A^-$$

The anion B^- of the eluent replaced with the analyte cation A^- bound to the positively charged ion X^+ on the surface of the stationary phase. The adsorption of the analyte to the stationary phase and desorption by the eluent ions is repeated during their journey in the column, resulting in the separation due to ion-exchange.

Molecules vary considerably in their charge properties and will exhibit different degrees of interaction with charged chromatography support according to differences in their overall charge, charge density and surface charge distribution. Net surface charge of all molecules with ionizable groups is highly pH dependent. Therefore pH of the mobile phase should be selected according to the net charge on a protein of interest within a mixture is opposite to that of matrix functional group, that it will displace the functional group counter ion and bind the matrix. On the other hand oppositely charged proteins will not be retained. Adsorbed protein analytes can be eluted by changing the mobile phase pH which effect the net charge of adsorbed protein, so its matrix binding capacity.

Moreover increasing the concentration of a similarly charged species within the mobile phase can be resulted in elution of bound proteins. During ion exchange chromatography for example in anion exchange as illustrated in Figure, negatively charged protein analytes can be competitively displaced by the addition of negatively charged ions. The affinity of interaction between the salt ions and the functional groups will eventually exceed that the interaction exists between the protein charges and the functional groups, resulting in protein displacement and elution by increasing gradually the salt concentration in the mobile phase.

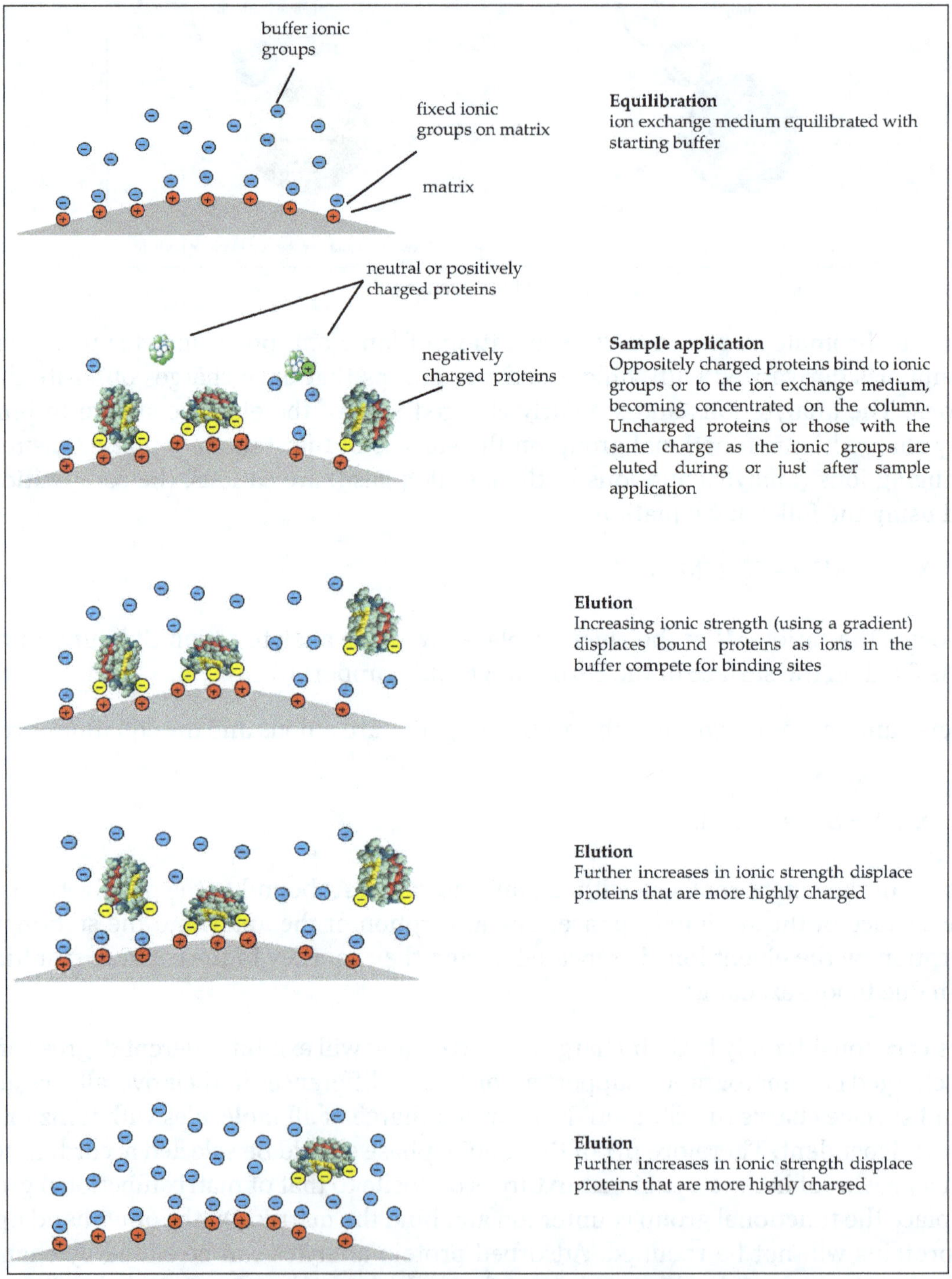

Separation steps in anion exchange chromatography (GE Healthcare).

Complex mixtures of anions or cations can usually be separated and quantitative amounts of each ion measured in a relatively short time by ion exchange chromatography. In classical ion-exchange chromatography separations have been performed in the open-column mode. Column which is loosely packed with stationary phase as small particles made of 1-2 cm diameter glass. The mobile phase or eluent contains the competing ion and is passed continuously into the column and percolates through it under gravity. Sample mixture is applied to the top of the column and allowed to pass into the bed of ion- exchange material. Eluent flow is then resumed and fractions of eluent are collected at regular intervals from the column outlet. Open column ion-exchange chromatography is very slow due to low eluent flow-rates. Increasing flow rate may result in deteriorated separation efficiency.

In modern ion-exchange chromatography the usage of high efficiency ion exchange materials combined with flow-through detection have overcome of these challenges. Separations are performed on the column which is filled with ion-exchanger as particles in uniform size. The particles of ion-exchange material are generally very much smaller than those used for classical open column ion-exchange chromatography. However ion-exchange resins used in modern chromatography have lower capacity than older resins. The eluent must be pumped through the column due to the small particle size of stationary phase. The sample mixture is applied into eluent by the injection port. Finally the separated ions are detected with a flow-through detection instrument.

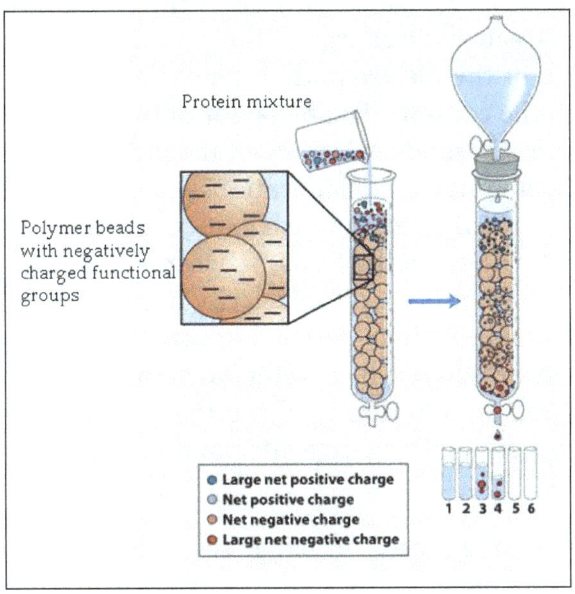

Ion exchange chromatography technique.

This technique has been used for the analyses of anions and cations, including metal ions, mono- and oligosaccharides, alditols and other polyhydroxy compounds, aminoglycosides (antibiotics), amino acids and peptides, organic acids, amines, alcohols, phenols, thiols, nucleotides and nucleosides and other polar molecules. It has been successfully applied to the analysis of raw materials, bulk active ingredients, counter ions, impurities, and degradation products, excipients, diluents and at different stages of the production process as well as for the analysis of production equipment cleaning solutions, waste streams, container compatibility and other applications. Wide applicability including high performance and high-throughput application formats, average cost, powerful resolving ability, large sample handling capacity and ease of scale-up as well as automation allow the ion exchange chromatography has become one of the most important and extensively used of all liquid chromatographic technique.

Although the extensive use of ion exchange chromatography the mechanism of the separation has not completely been elucidated. A considerable effort has been made to describe the ion exchange process theoretically. One of the important disadvantages of this technique is that this method provides no direct information on events occurring at the surface of the stationary phase, because the ion-exchange equilibrium is always determined by the balance between the solute interaction and the eluent interaction with the active sites of resin. Ion exchange is similar to sorption, since in both cases a solid takes up dissolved sample. The most important difference between them is in stoichiometric nature of ion exchange. Each ion removed from the solution is replaced by an equivalent amount of another ion of the same charge, while a solute is usually taken-up non-stoichiometrically without being replaced in sorption. Stoichiometric displacement based on the mass action law and describes the retention of a solute ion as an exchange process with the counter ion bound to the surface. According to this model, the retention of a protein under isocratic, linear conditions is related to counter ion concentration and can be represented by equilibrium as follow:

$$\log k = -(Z_P / Z_s) \log C_m + \log(\varphi Q)$$

k is the retention factor and C_m is the concentration of the counter ion in the mobile phase. Z_p/Z_s (= Z) is the ratio of the characteristic charge of the protein to the value of the counter ion and presents a statistical average of the electrostatic interactions of the protein with the stationary phase as it migrates through the column. The behavior of ion exchange chromatographic system can be explained by stoichiometric models. However, the mechanism of the ion exchange separation is more complex and stiochiometric consideration is inapplicable to long-range mechanisms, such as electrostatic interactions due to the distribution of ions in solution is also influenced by the electrostatic potential. Other interactions between solute-solute, solute-solvent and solvent-solvent also contribute to retention and selectivity in ion exchange. For example ion-dipole and dispersion interaction, should be included as important mechanisms. Additionally entropic contribution originating from solvent, such as water, structures around ion exchange sites should also be regarded as important. In addition to these the primary separation mechanism is the electrostatic interaction between ion-exchange sites and counter ions in ion exchange chromatography.

An important feature differentiating the ion exchange resins from other types of gels is the presence of functional groups. The groups are attached to the matrix. The ion exchange process between the ions in the solution takes place on these functional groups. The exchange of ions between the ion exchange resin and the solution is governed by two principles:

- The process is reversible, only rare exceptions are known.
- The exchange reactions take place on the basis of equivalency in accordance with the principle of electro neutrality. The number of milimoles of an ion sorbed by an exchange should correspond to the number of milimoles of an equally charged ion that has been released from the ion exchange.

Equilibrium is established for each sample component between the eluent and stationary phases when a sample is introduced into the ion-exchange chromatography. The distribution of component (A) between the two phases is expressed by the distribution coefficient, "D_A".

$$D_A = [A]_r / [A]_m$$

The value of D_A is dependent on the size of the population of molecules of component A in the stationary and eluent phases. As the equilibrium is dynamic, there is a continual, rapid interchange of molecules of component A between the two phases. The fraction of time, fm, that an average molecule of A spends in the mobile phase is given by:

- f_m = Amount of A in the mobile phase/Total amount of A.
- $f_m = [A]_m V_m/[A]_m V_m + [A]_r w = 1/1 + D_A (w/V_m)$.
- $k' = D_A (W/V_m)$.
- $f_m = 1/1 + k'$.
- w : Weight of the stationary phase.
- V_m: Volume of the mobile phase.

The mechanism of the anion and cation exchange are very similar. When analytes enter to the ion exchange column, firstly they bind to the oppositely charged ionic sites on the stationary phase through the Coulombic attraction. In accordance with Coulomb's law, the interactions between ions in the solute and oppositely charged ligands on the matrix in ion-exchange chromatography are due to the electrostatic forces. Coulomb's law is given by the equation as follow:

- $f = q_1q_2/\varepsilon r^2$.
- f: Interaction electrostatic force.
- q_1q_2: The charge on ions.
- ε: Dielectric constant of the medium.
- r: The distance between charges.

If the charges on both ions are same (both are positive or negative) the force is repulsive, if they are different (one positive and the other negative) the force is attractive. When the ion charge of the species increase (Divalent ion should interact more strongly than a monovalent ion) and when the dielectric constant decrease (Two oppositely charged molecules increased more strongly in an organic solvent than in water), the interactions increase. On the other hand the distance between the charges increases the interactions decrease. Additionally, other interactions, especially, van der Waals forces participate to the Coulombic forces.

Ion exchange chromatography, which is also known as adsorption chromatography, is a useful and popular method due to its:

- High capacity
- High resolving power
- Mild separation conditions

- Versatility and wide speared applicability
- Tendency to concentrate the sample
- Relatively low cost

General components of an ion-exchange chromatography are presented as below:

- A high pressure pump with pressure and flow indicator, to deliver the eluent.
- An injector for introducing the sample into the eluent stream and onto the column.
- A column, to separate the sample mixture into the individual components.
- An oven, optional.
- A detector, to measure the analyte peaks as eluent from the column.
- A data system for collecting and organizing the chromatograms and data.

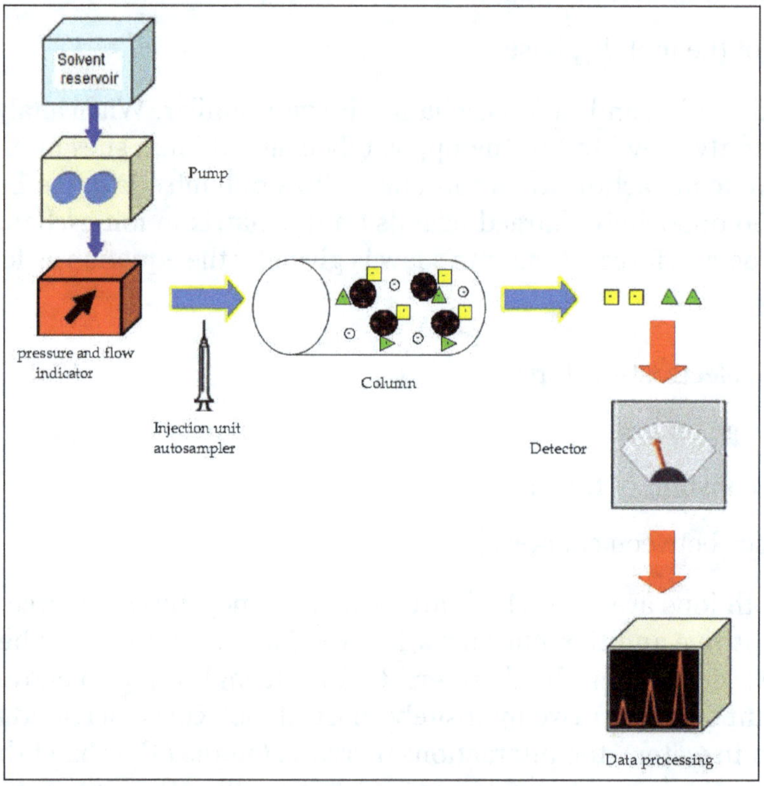

Ion-exchange Chromatography System.

In ion-exchange chromatography, adsorption and desorption processes are determined by the properties of the three interacting entities:

- The stationary phase.
- The constituents of the mobile phase.
- The solute.

Stationary Phase

Selection of a suitable ion-exchange matrix probably is the most important in ion exchange protocol and is based on various factors such as; ion exchanger charge/strength, linear flow rate/sample volume and sample properties. In ion-exchange chromatography, numerous stationary phases are available from different manufacturers, which vary significantly in a number of chemical and physical properties. Stationary phases comprised of two structural elements; the charged groups which are involved in the exchange process and the matrix on which the charged groups are fixed. Ion exchangers are characterized both by the nature of the ionic species comprising the fixed ion and by the nature of the insoluble ion-exchange matrix itself.

Ion exchangers are called cation exchangers if they have negatively charged functional groups and possess exchangeable cations. Anion exchangers carry anions because of the positive charge of their fixed groups. The charged groups determine the specificity and strength of protein binding by their polarity and density while the matrix determines the physical and chemical stability and the flow characteristics of the stationary phase and may be responsible for unspecific binding effects.

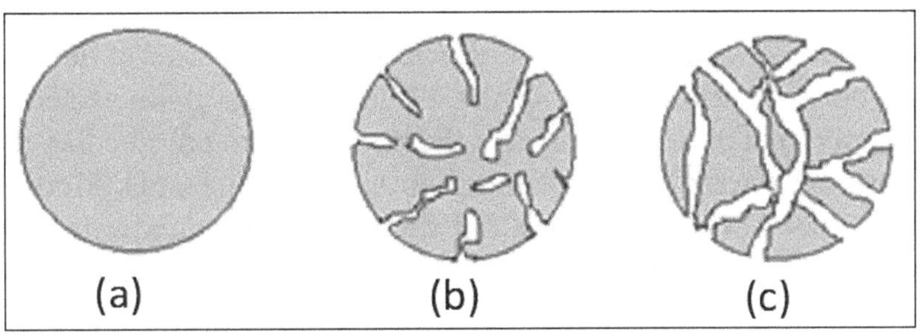

Schematic presentation of different matrix types (a) non-porous beads (b) microporous beads (c) macroporous beads.

General structure (fibrous or beaded form), particle size and variation, pore structures and dimensions, surface chemistry (hydrophilic or hydrophobic), swelling characteristics of matrix are important factors which effect chromatographic resolution. Porosity of ion exchange beads can be categorized as non-porous, microporous and macroporous. High porosity offers a large surface area covered by charged groups and so provides a high binding capacity. However when compared with beaded matrix fibrous ion exchangers based on cellulose exhibit lower chromatographic resolution. On the other hand high porosity is an advantage when separating large molecules and pre-fractionation. Non-porous matrices are preferable for high resolution separations when diffusion effects must be avoided. Micropores increase the binding capacity but cause to a band broadening. Another disadvantage of microporous beads is that protein can bind to the surface of the beads near to the pores, so penetration of proteins into the pores can prevent or slow down. These problems are overcome by using macroporous particles with pore diameters of about 600-800 nm which are introduced recently. These kinds of particles behave differently compared to microporous materials with respect to microflow characteristics the new term perfusion chromatography has been created.

Furthermore a new matrix type which has been recently introduced is based on a completely new principle and exhibits improved chromatographic features when compared with conventional ion exchangers. This matrix which is known as continuous bed does not consist of ion exchange beads or fibers. The matrix is synthesized in the column by polymerization and established from continuous porous support consisting of a nodule chains. The advantages of that matrix are mainly due to the more homogeneous mobile phase flow and short diffusion distances for the proteins. This is explained by the non-beaded form and the unique pore structure of the support.

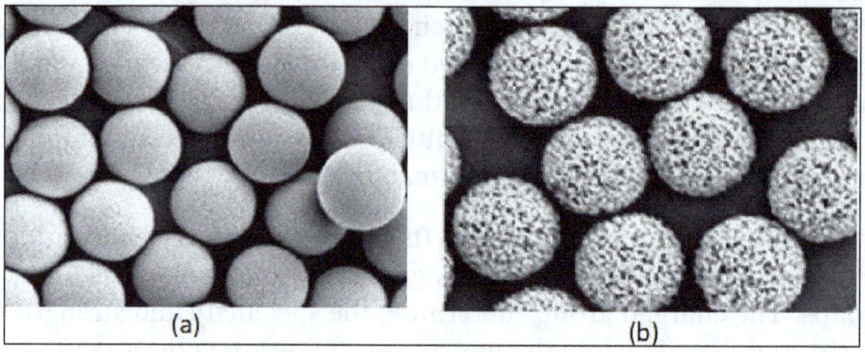

(a) Non-porous beads (b) Porous beads.

Size, size distribution and porosity of the matrix particles are the main factors which affect the flow characteristics and chromatographic resolution. Small particles improved chromatographic resolution. Stationary phases with particle of uniform size are superior to heterogeneous materials with respect to resolution and attainable flow rates. The pore size of ion exchange bead directly affects the binding capacity for a particular protein dependent on the molecular weight of the protein because it determines the access of proteins to the interior of the beads. Binding of large proteins can be restricted to the bead surface only so that the total binding capacity of the ion exchanger is not exploited Pore diameter of 30 nm is optimal for proteins up to a molecular weight of about 200.000 Da.

Continuous bed matrix.

In order to minimize non-specific interactions with sample components inert matrix should be used. High physical stability provides that the volume of the packed medium remains constant despite extreme changes in salt concentration or pH for improving reproducibility and avoiding the need to repack columns. High physical stability and uniformity of particle size facilitate high flow rates, particularly during cleaning or re-equilibration steps, to improve throughput and productivity. There are pH and pressure limits for each stationary phase. For example pH values higher than 8 should not use in silica based materials which are not coated with organic materials. Matrix stability also should be considered when the chemicals such as organic solvents or oxidizing agents should be required to use or when they are chosen for column cleaning.

Matrices which are obtained by polymerization of polystyrene with varying amounts of divinylbenzene are known as the original matrices for ion exchange chromatography. However these matrices have very hydrophobic surface and proteins are irreversibly damaged due to strong binding. Ion exchangers which are based on cellulose with hydrophilic backbones are more suitable matrices for protein separations. Other ion exchange matrices with hydrophilic properties are based on agarose or dextran. Several matrix types and their important properties can be listed as follow:

Matrix Materials

- Cellulose: Hydrophilic surface, enhanced stability by cross-linking, inexpensive.
- Dextran: Considerable swelling as a function of ionic milieu, improved materials by cross-linking).
- Agarose: Swelling is almost independent of ionic strength and pH, high binding capacity obtained by production of highly porous particles.
- Polyacrylamide: Swelling behavior similar to dextran.
- Acrylate-copolymer: High pH stability.
- Polystyrene-divinilybenzene: Hydrophobic surface, low binding capacity for proteins.
- Coated polystyrene-divinilybenzene: Hydrophilic surface.
- Silica: Unstable at pH > 8, rigid particles.
- Coated Silica: Hydrophilic surface.

In addition to electrostatic interactions between stationary phase and proteins, some further mechanisms such as hydrophobic interactions, hydrogen bonding may contribute to protein binding. Hydrophobic interactions especially occur with synthetic resin ion exchangers such as which are produced by copolymerization of styrene and divinylbenzene. These materials are not usually used for separation of proteins. However new ion exchange materials that consist of styrene-divinylbenzene copolymer beads coated with hydrophilic ion exchanger film were introduced. According to the retention behavior of some proteins, it is considered that coating of the beads so efficient that unspecific binding due to hydrophobic interactions cannot be observed. Silica particles have also been coated with hydrophilic matrix. Acrylic acid polymers are also used for the protein separation in ion exchange chromatography. These polymers are especially suitable for purification of basic proteins.

The functional groups substituted onto a chromatographic matrix determine the charge of an ion exchange medium; positively-charged anion exchanger or a negatively-charged cation exchanger. Both exchangers can be further classified as strong and weak type as shown in Table. The terms weak and strong are not related to the binding strength of a protein to the ion exchanger but describe the degree of its ionization as a function of pH. Strong ion exchangers are completely ionized over a wide pH range, while weak ion exchangers are only partially ionized a narrow pH range.

Therefore with strong ion exchangers proteins can adsorb to several exchanger sites. For this reason strong ion exchangers are generally used for initial development and optimization of purification protocols. On the other hand weak ion exchangers are more flexible in terms of selectivity and are a more general option for the separation of proteins that retain their functionality over the pH 6-9 range as well as for unstable proteins that may require mild elution conditions. Alkylated amino groups for anion exchangers and carboxy, sulfo as well as phosphato groups for cation exchangers are the most common functional groups used on ion exchange chromatography supports. Sulfonic acid exchangers are known as strong acid type cation exchangers. Quaternary amine functional groups are the strong base exchangers whereas less substituted amines known as weak base exchangers. Number and kind of the substituents are determined the basicity of amino-groups. Immobilized tertiary and quaternary amines proved to be useful for ion exchange chromatography. Immobilized diethylaminoethyl and carboxymethyl groups are the most widely used ion exchangers.

The ion exchange capacity of an ion-exchanger is determined by the number of functional groups per unit weight of the resin. The total ionic capacity is the number of charged functional groups per ml medium, a fixed parameter of each medium and can be given as mval/ml for small ions. Density and accessibility of these charged groups both on the surface and within the pores define the total binding capacity. Ionic medium and the presence of other proteins if a particular protein is considered also affect the binding capacity. However, under defined conditions, the amount of the certain protein which is bound to ion exchanger is more suitable parameter for determining and comparing the capacity of ion exchange chromatography. Albumin for anion exchangers and hemoglobin for cation exchangers is usually used for this purpose. Determination of the binding capacity before the experiment is generally recommended because the capacity for a particular protein depends on its size and also on the sample composition. The binding capacity of a column can be increased for proteins which are retained on the column at high salt concentrations. The salt concentration is adjusted to a suitable concentration in which the protein of interest tightly bound to the ion exchanger while others which have lower affinity pass through the column without occupying binding sites.

Table: Weak and Strong type anion and cation exchangers.

Exchange Type	Ion exchange group	Buffer counter ions	pH range	Commercial samples
Strong cation	Sulfonic acid (SP)	Na^+, H^+, Li^+	4-13	Capto®S
				SP Sepharose®
				SP Sephadex®
				TSKgel SP_5PW
Weak cation	Carboxylic acid	Na^+, H^+, Li^+	6-10	CM Cellulose
				CM Sepharose®
				CM Sephadex®
				CM Sepharose® CL6B
				TSKgel CM$_{-5}$PW

Strong anion	Quaternary amine (Q)	Cl^-, $HCOO_3^-$, CH_3COO^-, SO_4^{2-}	2-12	Q Sepharose® Capto®Q Dowex®1X2 Amberlite® / Amberjet® QAE Sephadex®
Weak anion	Primary amine Secondary amine Tertiary amine (DEAE)	Cl^-, $HCOO_3^-$, CH_3COO^-, SO_4^{2-}	2-9	DEAE-Sepharose® Capto® DEAE DEAE Cellulose

Mobile Phase (Eluent)

In ion exchange chromatography generally eluents which consist of an aqueous solution of a suitable salt or mixtures of salts with a small percentage of an organic solvent are used in which most of the ionic compounds are dissolved better than in others in. Therefore the application of various samples is much easier. Sodium chloride is probably the most widely used and mild eluent for protein separation due to has no important effect on protein structure. However NaCl is not always the best eluent for protein separation. Retention times, peak widths of eluted protein, so chromatographic resolution are affected by the nature of anions and cations used. These effects can be observed more clearly with anion exchangers as compared to cation exchangers. The salt mixture can itself be a buffer or a separate buffer can be added to the eluent if required. The competing ion which has the function of eluting sample components through the column within reasonable time is the essential component of eluting sample. Nature and concentration of the competing ions and pH of the eluent are the most important properties affecting the elution characteristics of solute ions.

The eluent pH has considerable effects on the functional group which exist on the ion exchange matrix and also on the forms of both eluent and solute ions. The selectivity coefficient existing between the competing ion and a particular solute ion will determine the degree of that which competing ion can displace the solute ion from the stationary phase. As different competing ions will have different selectivity coefficients, it follows that the nature of competing ion will be an important factor in determining whether solute ions will be eluted readily. The concentration of competing ion exerts a significant effect by influencing the position of the equilibrium point for ion-exchange equilibrium. The higher concentration of the competing ion in the eluent is more effectively displace solute ions from the stationary phase, therefore solute is eluted more rapidly from the column. Additionally elution of the solute is influenced by the eluent flow-rate and the temperature. Faster flow rates cause to lower elution volumes because the solute ions have less opportunity to interact with the fixed ions. Temperature has relatively less impact, which can be change according to ion exchange material type. Enhancement of the temperature increases the rate of diffusion within the ion-exchange matrix, generally leading to increased interaction with the fixed ions and therefore larger elution volumes. At higher temperatures chromatographic efficiency is usually improved.

Eluent degassing is important due to trap in the check valve causing the prime loose of pump. Loss of prime results in erratic eluent flow or no flow at all. Sometimes only one pump head will lose its prime and the pressure will fluctuate in rhythm with the pump stroke. Another reason for removing dissolved air from the eluent is because air can get result in changes in the effective concentration of the eluent. Carbon dioxide from air dissolved in water forms of carbonic acid. Carbonic acid can change the effective concentration of a basic eluent including solutions of sodium hydroxide, bicarbonate and carbonate. Usually degassed water is used to prepare eluents and efforts should be made to keep exposure of eluent to air to a minimum after preparation. Modern inline degassers are becoming quite popular.

For separation the eluent is pumped through the system until equilibrium is reached, as evidenced by a stable baseline. The time required for equilibrium may vary from a couple of minutes to an hour or longer, depending on the type of resin and eluent used. Before the sample injection to the column should be equilibrated with eluent to cover all the exchange sites on the stationary phase with the same counter ion. When the column is equilibrated with a solution of competing ion, counter ions associated with the fixed ions being completely replaced with competing ions. In this condition the competing ions become the new counter ions at the ion exchange sites and the column is in the form of that particular ion.

Isocratic elution or gradient elution can be applied for elution procedure. A single buffer is used throughout the entire separation in isocratic elution. Sample components are loosely adsorbed to the column matrix. As each protein will have different distribution coefficient separation will achieved by its relative speeds of migration over the column. Therefore in order to obtain optimum resolution of sample components, a small sample volume and long exchanger column are necessary. This technique is time consuming and the desired protein invariably elutes in a large volume. However no gradient-forming apparatus is required and the column regeneration is needless. Alteration in the eluent composition is needed to achieve desorption of desired protein completely. To promote desired protein desorption continuous or stepwise variations in the ionic strength and/or pH of the eluent are provided with gradient elution. Continuous gradients generally give better resolution than stepwise gradients.

Additives which are protective agents found in the mobile phase are generally used for maintain structure and function of the proteins to be purified. This is achieved by stimulating an adequate microenvironment protection against oxidation or against enzymatic attacks. Any additives used in ion exchange chromatography, should be checked for their charge properties at the working pH in order to avoid undesired effects due to adsorption and desorption processes during chromatography. It is recommended to include in the elution buffer those additives in a suitable concentration which have been used for stabilization and solubilization of the sample. Otherwise precipitation may occur on the column during elution. For example; zwitterionic additives such as betaine can prevent precipitation and can be used at high concentrations without interfering with the gradient elution. Detergents are generally useful for solubilization of proteins with low aqueous solubility. Anionic, cationic, zwitterionic and non-ionic (neutral) detergents can be used during ion exchange chromatography. Guanidine hydrochloride or urea, known as denaturing agents can be used for initial solubilization of a sample and during separation. However, they should use if there is a requirement. Guanidine is a charged molecule and therefore can participate to the ion exchange process in the same way as NaCl during separation process. Commonly used eluent additives which have been successfully used in ion exchange chromatography can be given as follow:

- EDTA: Ethylenediamine tetraacetic acid,
- Polyols: Glycerol, glucose, and saccharose,
- Detergents,

- Urea and guanidinium chloride,
- Lipids,
- Organic solvents,
- Zwitterions,
- Sulfhydryl reagents,
- Ligands,
- Protease inhibitors.

Buffer

In ion exchange chromatography, pH value is an important parameter for separation and can be controlled and adjusted carefully by means of buffer substances. In order to prevent variation in matrix and protein net charge, maintenance of a constant mobile phase pH during separation is essential to avoid pH changing which can occur when both protein and exchanger ions are released into the mobile phase. By means of buffer substances pH value can be controlled and adjusted. Concentration of H^+ and the buffering component influence the protein binding to the stationary phase, chromatographic resolution and structural as well as functional integrity of the protein to be separated. Thus a suitable pH range, in which the stability of sample is guaranteed, has to be identified. Keeping of the sample function is related with the preservation of its three dimensional structure as well as with its biological activity. A number of buffers are suitable for ion-exchange chromatography.

A number of important factors influences the selection of mobile phase including buffer charge, buffer strength and buffer pH. Properties of good buffers are high buffering capacity at the working pH, high solubility, high purity and low cost. The buffer salt should also provide a high buffering capacity without contributing much to the conductivity and should not interact with the ion exchanger functional groups as well as with media. The buffering component should not interact with the ion exchanger because otherwise local pH shifts can occur during the exchange process which may interfere the elution. Interactions with stationary phase as well as with additives of the mobile phase and with subsequent procedures may be occur with buffer component and selected pH range. Precipitation of the mobile phase components can be observed for example when phosphate buffer and several di- and trivalent metal ions such as Mg^{+2} and Ca^{+2} are mixed or when anionic detergents (i.e. cholate) are used under acidic conditions or in the presence of multivalent metal ions. Precipitation of metal oxides and hydroxides can occur under alkaline conditions. Buffer components may also affect enzymatic assays used for screening and analysis of chromatography fractions. The concentration of buffer salts usually ranges from 10 to 50 mM. Commonly used buffers are presented in Table for cation and anion exchange chromatography.

Table: Commonly used buffers for cation-exchange chromatography.

Substance	pKa	Working pH
Citric acid	3.1	2.6-3.6
Lactic acid	3.8	3.4-4.3
Acetic acid	4.74	4.3-5.2

2-(*N*-morpholino)ethanesulfonic acid	6.1	5.6-6.6
N-(2-acetamido)-2-iminodiacetic acid	6.6	6.1-7.1
3-(*N*-morpholino)propanesulfonic acid	7.2	6.7-7.7
Phosphate	7.2	6.8-7.6
N-(2-hydroxyethyl)piperazine-*N'*-(2-ethanesulfonic acid)	7.5	7.0-8.0
N,N-bis(2-hydroxyethyl)glycine	8.3	7.6-9.0

Generally, applications of ion exchange chromatography are performed under slightly acidic or alkali conditions, pH range 6.0-8.5 but there are also more acidic and more alkali buffers. Additionally the buffering component should not act as an eluting ion by binding to the ion exchanger. Anionic buffer component such as phosphate or MOPS in cation exchange chromatography and cationic buffers such as ethanolamine, Tris and Tricine in anion exchange chromatography are recommended. Besides interactions of buffer component with stationary phase, there are also possible interactions with additives of the mobile phase. To achieve sufficient buffer capacity the pKa of the buffer component should be as close to the desired pH value as possible difference no more than ± 0.5 pH units. However there are examples of successful separations at which the buffering capacity is very low. It has to be considered that the pKa is a temperature dependent value. Performing on ion exchange separation with the same elution buffer at room temperature or in the cold room can have a remarkable effect on the buffer capacity. For optimal binding of a sample ion to an ion-exchanger the ionic strength and thus also the buffer concentrations has to be low in sample and equilibration buffers.

Table: Commonly used buffers for anion-exchange chromatography.

Substance	pKa	Working pH
N-Methyl-piperazine	4.75	4.25-5.25
Piperazine	5.68	5.2-6.2
Bis-Tris	6.5	6.0-7.0
Bis-Tris propane	6.8	6.3-7.3
Triethanolamine	7.8	7.25-8.25
Tris	8.1	7.6-8.6
N-Methyl-diethanolamine	8.5	8.0-9.0
Diethanolamine	8.9	8.4-9.4
Ethanolamine	9.5	9.0-10.0
1,3-Diaminopropane	10.5	10.0-11.0

Detection

Conductivity detector is the most common and useful detector in ion exchange chromatography. However UV and other detectors can also be useful. Conductivity detection gives excellent sensitivity when the conductance of the eluted solute ion is measured in an eluent of low background conductance. Therefore when conductivity detection is used dilute eluents should be preferred and in order for such eluents, to act as effective competing ions, the ion exchange capacity of the column should be low.

Although recorders and integrators are used in some older systems, generally in modern ion exchange chromatography results are stored in computer. Retention time and peak areas are the most useful information. Retention times are used to confirm the identity of the unknown peak by comparison with a standard. In order to calculate analyte concentration peak areas are compared with the standards which is in known concentration.

Direct detection of anions is possible, providing a detector is available that responds to some property of the sample ions. For example anions that absorb in the UV spectral region can be detected spectrophotometrically. In this case, an eluent anion is selected that does not absorb UV. The eluent used in anion chromatography contains an eluent anion, E^-. Anions with little or no absorbance in the UV spectral region can be detected spectrophotometrically by choosing a strongly absorbing eluent anion. An anion with benzene ring would be suitable. Usually Na^+ or H^+ will be the cation associated with E^-. The eluent anion must be compatible with the detection method used. For conductivity the detection E should have either a significantly lower conductivity than the sample ions or be capable of being converted to a non-ionic form by a chemical suppression system. When a spectrophotometric detection is employed, E will often be chosen for its ability to absorb strongly in the UV or visible spectral region. The concentration of E^- in the eluent will depend on the properties of the ion exchanger used and on the types of anions to be separated.

Applications of Ion Exchange Chromatography

- It is extremely used in the analysis of amino acids. The amino acid "Autoanalyzer" is based on in exchange principle.
- To determine the base composition of nucleic acids. Chargaff used this technique for established the equivalence of Adenine and Thymine; Guanine and Cytosine.
- This is the most effective method for water purification. Complete deionization of water (or) a non-electrolyte solution is performed by exchanging solute cations for hydrogen ions and solute anions for hydroxyl ions. This is usually achieved by the method is used for the softening of drinking water.
- Proteins are also successfully separated by this technique.
- It is also used for the separation of many vitamins, other biological amines, and organic acids and bases.

References

- What-is-chromatography: lab-training.com, Retrieved 01, May 2020
- Gas-chromatography: microbenotes.com, Retrieved 21, January 2020
- Affinity-chromatography: microbenotes.com, Retrieved 15, July 2020
- Ion-exchange-chromatography-and-its-applications, column-chromatography: intechopen.com, Retrieved 09, March 2020
- Applications-of-Ion-Exchange-Chromatography: biochemden.com, Retrieved 20, June 2020

3

Spectroscopic Methods

Spectroscopy refers to the study of absorption and emission of radiations by matter. The various types of spectroscopy are mass spectrometry, molecular spectroscopy, atomic absorption spectroscopy, atomic emission spectroscopy, flame emission spectroscopy and Raman spectroscopy. This chapter has been carefully written to provide an easy understanding of these types of spectroscopy.

Spectrochemical analysis is the methods of chemical analysis that depend upon the measurement of the wavelength and the intensity of electromagnetic radiation. Its major use is in the determination of the arrangement of atoms and electrons in molecules of chemical compounds on the basis of the amounts of energy absorbed during changes in the structure or motion of the molecules.

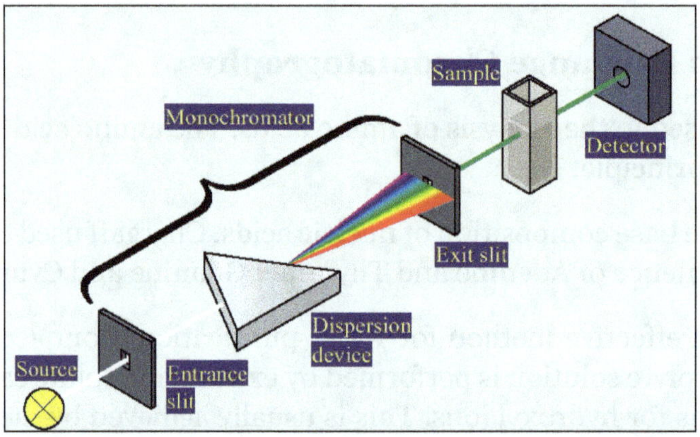

A spectrograph separates light into its constituent wavelengths.
The spectroscopy of optical light produces the familiar rainbow.

Spectroscopy is the technique of splitting light (or more precisely electromagnetic radiation) into its constituent wavelengths (a spectrum), in much the same way as a prism splits light into a rainbow of colours. However, in general, a spectrum is generally more than a simple 'rainbow' of colours. The energy levels of electrons in atoms and molecules are quantised, and the absorption and emission of electromagnetic radiation only occurs at specific wavelengths. Consequently, spectra are not smooth but punctuated by 'lines' of absorption or emission.

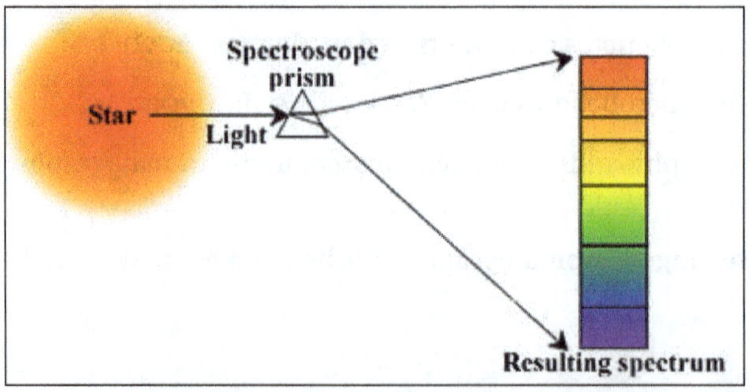

The spectra of stars are not simple smooth 'rainbows' of colours, Rather they are punctuated by absorption and emission lines.

Old style spectroscopy was carried out using a prism and photographic plates. These days, modern spectroscopy uses diffraction gratings to disperse the light, which is then projected onto CCDs (Charge Coupled Devices) similar to those used in digital cameras. The 2-dimensional spectra are easily extracted from this digital format and manipulated to produce 1-dimensional spectra like the galaxy spectrum shown below. These 1-dimensional spectra contain an astonishing amount of useful data:

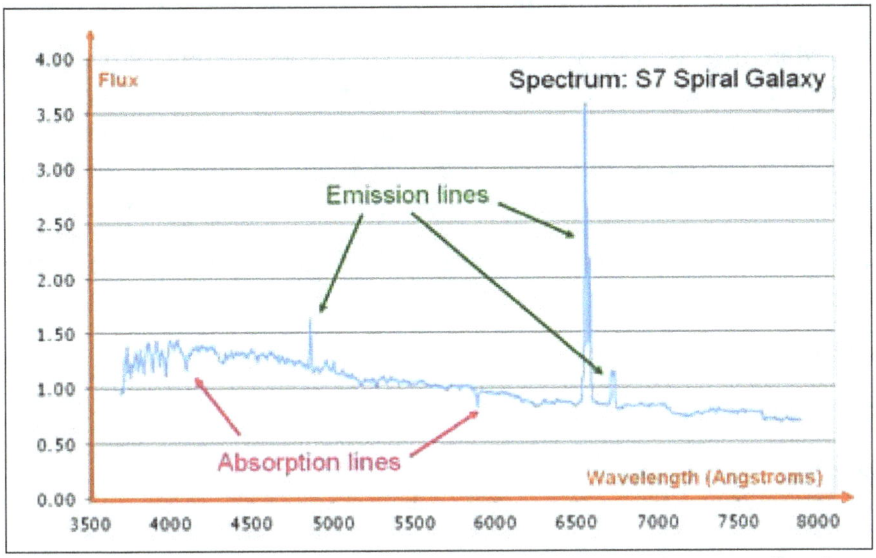

The spectrum of an S7 spiral galaxy showing both emission and absorption line features. Wavelength is measured in angstroms while the flux is in arbitrary units.

- The precise position (wavelength) at which known emission and absorption lines are detected can be used to measure the redshift of the observed object. For example, if the Hβ (486.2 nm) hydrogen line is detected at 487.8 nm, we calculate that the object has a recession velocity of 1,000 km/sec. Apart from measuring distances, this type of analysis can also be used to detect spectroscopic binaries and extrasolar planets.

- The widths of both the absorption and emission lines can be used to measure the internal velocity dispersion of complex objects (e.g. the average velocity of stars within galaxies). This is achieved through measurement of spectral line broadening from which the velocity broadening can be calculated.

- Many (indeed most) absorption and emission lines are produced by metals. The depths or heights of these lines can be used to estimate the abundances of the metals responsible. In stars, these also permit the measurement of the temperature and pressure of the stellar atmosphere. In objects such as galaxies, they permit an estimation of age, though the spectra are somewhat susceptible to the age-metallicity degeneracy making this a difficult procedure.

- Finally, the overall shape of the spectrum contains clues to many other aspects of the observed population including age and interstellar reddening (extinction).

The enormous quantities of information contained within a single spectrum makes spectroscopy one of the most powerful tools at an astronomer's disposal.

Mass Spectrometry

Mass spectrometry is an analytical method useful for calculating the mass-to – charge ratio (m/z) of one or more molecules in the sample. Such measurements may also often be used to determine the precise molecular weight of the sample components. Mass spectrometry is an analytical method to find the molecular mass of a compound and indirectly helped to prove the identity of isotopes.

- Principle of Mass Spectrometry: Based on Newton's second law of motion and momentum, a mass spectrometer uses this property of matter to plot ions of varying masses on a mass spectrum. From the law, we infer how much mass is relevant to the inertia and acceleration of a body. This principle is applied to the aspect where ions with different mass to charge ratios are deflected by different angles in an electric or magnetic field.
- Mass Spectrum: A mass spectrum is a graph obtained by performing mass spectrometry. It is a relation between the mass to charge ratio and ion signal.
- Mass Spectrometry Diagram:
    - Inlet system
    - Ionization
    - Deflector
    - Ion detector

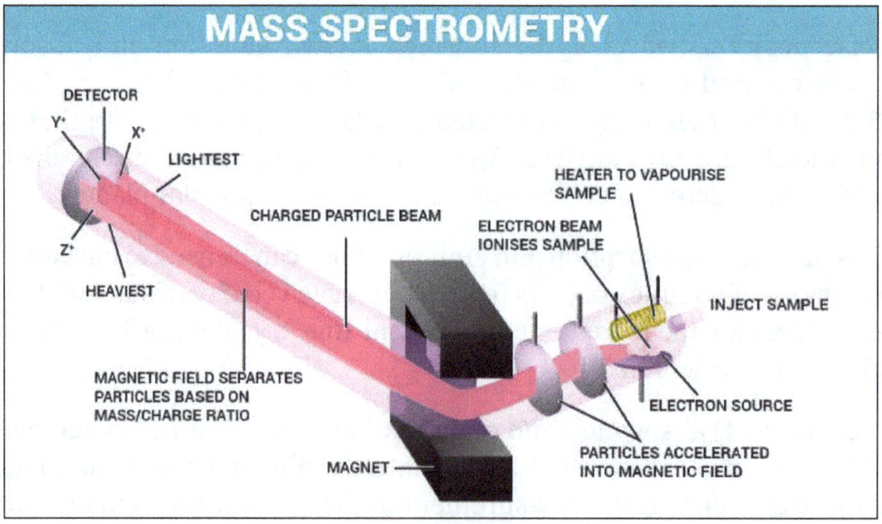

Mass Spectrometry Instrumentation.

Mass Spectrometry Detectors

At different deflections a detector counts the number of ions. The data are plotted as a graph or continuum of various masses. Detectors function by recording the induced charge or current generated by an ion hitting or passing through a surface. Since the signal is very small it is possible to use an electron amplifier, Faraday cup, or ion-to-photon detector. To generate a spectrum the signal is greatly amplified.

The Mass Analyzer

When ionized, the ions are sorted and divided according to the mass-to – charge (m/z) ratio. A variety of mass analyzers are currently available, each of which has trade-offs related to speed of operation, separation resolution and other technical criteria. The different forms in use at the Broad Institute are described in the following section. The mass analyzer often works in concert with the ion detection system.

Quadrupole

Mass spectrometry determines the chemical by calculating the typical mass fragments formed by the ionization of the material. Test molecules are ionized by an electron beam, and the resulting molecular ion and component ions travel into a mass analyzer where their masses are measured. Mass spectrometry is generally considered the benchmark for identification of unknown organic chemicals because it is highly sensitive and selective, and mass spectra are easily searchable against vast reference databases.

Mass Spectrometry Instrumentation

The four main parts of mass spectrometry are discussed below:

- Ionizer: The bombarding of the sample is done by the electrons. These electrons move between cathode and anode. When the sample passes through the electron stream between the cathode and anode, electrons with high energy knock electrons out of the sample and form ions.
- Accelerator: The ions placed between a set of charged parallel plates get attracted to one plate and repel from the other plate. The acceleration speed can be controlled by adjusting the charge on the plates.
- Deflector: Magnetic field deflects ions based on its charge and mass. If an ion is heavy or has two or more positive charges, then it is least deflected. If an ion is light or has one positive charge, then it is deflected the most.
- Detector: The ions with correct charge and mass move to the detector. The ratio of mass to charge is analyzed through the ion that hits the detector.

Working Principle of a Mass Spectrometer

In a regular mass spectrometer, we initially have the material to be analyzed, but we need it to be ionized to pass through the spectrometer with enough energy. Thus, the sample is bombarded by electrons to ionize it. This ionized beam is now passed through a series of electric or magnetic fields depending on the type of the sample and its properties. The ions are deflected by the field through which they are passed through in such a way that the ions with the same mass to signal ratio will follow the same path to the detector.

These charged and deflected ions are now incident onto a detector which is capable of distinguishing the charged particles falling on it. Based on the mass spectrum produced by the charged ions, we can identify the atoms or molecules constituting the sample by comparing them with known masses or through a characteristic fragmentation pattern.

Applications of Mass Spectrometry

Mass spectrometry is an efficient method to elucidate the chemical composition of a sample or molecule. More recently, it has been used to classify biological products, in particular proteins and protein complexes, in a number of species. Usually, mass spectrometers can be used to classify unknown substances by molecular weight measurement, to measure known compounds, and to determine the structure and chemical properties of molecules.

- Due to its capability to distinguish between substances, Mass spectrometry is used to determine unknown substances.
- To identify the isotopes of a substance.
- In analytical laboratories that study the chemical, physical and biological properties of substances. It is favored over several other analytical techniques as it has less background interference since it is performed in a vacuum.

Fields of Application of Mass Spectrometry

Key application and field of application	Explanation
Elemental and isotopic analysis: • Physics • Radiochemistry • Geochemistry	Elemental identification and isotopic abundance measurement of both short lived and stable species in physical and radioactivity in geochemistry and more recently in the life sciences.
Organic and bio-organic analysis: • Organic chemistry • Polymer chemistry • Biochemistry and medicine	Identification and structural characterization of molecules from small to very large as provided either by chemistry, physiological processes or polymer chemistry.
Structural elucidation: • Organic chemistry • Polymer chemistry • Biochemistry and medicine	Mass spectrometric experiments can be arranged consecutively to study mass selected ions in tandem mass spectrometry. Eventually products are subjected to a third level and so forth.
Characterization of ionic species and chemical reactions: • Physical chemistry • Thermochemistry	Tandem MS provide as elegant means for the study of unimolecular or bimolecular reactions of gas phase ions and for the determination of ion energetics.
Mass Spectral Imaging: • Biomedical studies • Material sciences	Mass spectra can be obtained from micrometer sized areas on surfaces, translating the lateral distribution of compounds on surfaces into images, which in turn can be correlated to optical images.

Mass Spectrometry Advantages and Disadvantages

Mass spectrometry is used for both qualitative and quantitative study of chemical substances.

These can be used to classify a sample's elements and isotopes, to determine molecular masses, and as a tool for helping to classify chemical structures. This can calculate the purity of the samples and the molar mass.

A big advantage of mass spec is that it is incredibly sensitive (parts per million) over many other techniques. It is an excellent tool for identifying or confirming presence of unknown components in a sample. The disadvantages of mass spec are that identifying hydrocarbons that produce similar ions is not very good and it is not able to separate optical and geometric isomers. The disadvantages are offset by combining MS with other methods, for example gas chromatography. One disadvantage of mass spectrometry is over hydrocarbons. The method fails to distinguish between hydrocarbons producing similarly fragmented ions.

Atomic Absorption Spectroscopy

Atomic Absorption Spectroscopy is an instrumental analysis technique for rapid trace metal analysis. It is based on element specific wavelength light absorption by ground state atoms in the flame or electro thermal graphite furnace. It finds immense applications in the analysis for trace metals in soils, lakes, rivers, oceans, and drinking water, pharmaceuticals, foods and beverages, geological and mineralogical samples, petroleum products, biological fluids and specimens and forensic analysis. It is common to get results in ppm levels and a higher sensitivity of ppb levels when we using graphite furnace atomisation.

Uses of Atomic Absorption Spectroscopy

Atomic Absorption Spectroscopy provides cost-effective viable solutions for the analysis of trace amounts of metals in the entire range of natural and manmade materials such as Geological samples, Environmental samples, Biological Specimens, Agricultural produce and soils, Pharmaceuticals, Foods and Drinking water. The technique affords advantages of speed, sensitivity and precision over the classical gravimetric methods. Introduction of accessories such as graphite furnace, flow injection analysis and improvements in the suppression of matrix interferences have further contributed to improvement in sensitivity and selectivity of analytes in complex matrices.

Atomic Absorption Spectroscopy applications in the field of environment, drinking water, mining and mineralogy, oceanographic studies, soils, pharmaceuticals, foods, toys, forensic investigations are of great significance. The list is endless and presence or absence of trace metals is a factor that cannot be overlooked for evaluation of characteristics of materials or concerns regarding human health and safety.

The chemical techniques used for the analysis of trace metals have evolved from simple gravimetric methods to highly sophisticated time saving instrumental techniques. Atomic Absorption Spectroscopy is a popular technique which involves moderate investment and affordable operational cost. These features coupled with a high degree of accuracy and precision of results has contributed to the widespread presence of atomic absorption spectrometers in college laboratories, industrial laboratories and regulatory body laboratories across the world.

Principle of Atomic Absorption Spectroscopy

Atomic absorption spectroscopy (AAS) is based upon the principle that free atoms in the ground state can absorb light of a certain wavelength. These very specific wavelengths give the technique excellent specificity and detection limits in the AAS analysis. Absorption for each element is specific; no other elements absorb this wavelength. Typical applications of AAS include:

- Quantitative metal concentrations in solution.
- Analysis of lead in paint.
- Monitoring of trace metals in industrial effluent streams.
- Trace elements in product/raw materials along with ICP-MS.
- Analysis of additives and purity in steels and other metal alloys.
- Analysis of low-level contaminants.

Several analytical techniques have been applied for detection and quantitative estimation of trace metals in different types of matrices. Classical techniques based on gravimetric and titrimetry provided good accuracy but were time-consuming.

Atomic Emission Spectroscopy

In emission spectroscopy, a sample gets excited by absorption of thermal or electric energy and the emitted radiation is analysed for both qualitative and quantitative purposes. Emission spectroscopy is related to atoms. Liquid, solids are usually studied whereas analysis of gaseous samples is done rarely.

Principle

The complete spectrum is divided into various regions according to wavelength which are as follows:

- Ultraviolet region: Radiations of wavelengths up to 4000 Å.
- Visible region: Radiations of wavelengths between 4000 and 7300 Å.
- Infrared region: Radiations of wavelengths between 7300 and 35,000 Å.

Emission Spectra

A substance emits light when heated to a high temperature by thermal or electric energy. The emitted light after passing from monochromatic is allowed to enter the detecting system and then a spectrum is obtained which is known as emission spectrum. Emission spectra are of following types:

- Continuous spectra: This spectrum shows uninterrupted emission over a considerable wavelength region. Sharp lines or discrete bands are absent.

- Band spectra: A band is emitted by each molecule which is characteristic of that particular molecule. Band spectra are also known as molecular spectra.
- Line spectra: It consists of discrete irregularly lines.

Requirements for Excitation of Elements

To excite any element for emission of a single spectral line, energy equivalent to excitation potential of that element must be absorbed. A complete spectrum is obtained when energy equivalent to ionisation potential is absorbed. By employing an enough energetic source, emission spectra of different elements in the periodic table can be obtained. Emission spectroscopy is limited to metals and metalloids and not non-metals as spectral lines fall in the vacuum-ultraviolet region and their study is very difficult.

The energy from an excitation source is fixed and therefore this energy is divided up among thousands of lines and thus affects sensitivity of each line. Therefore, it is not easy to study the complex spectra of elements like uranium, etc.

Advantages of Emission Spectroscopy Over Flame Photometry

Emission spectroscopy has many advantages over flame photometry. Some of them are:

- Universal in application.
- Can make use of transitions between upper excited states and measure lines which are irreversible at high concentrations.
- High sensitivity.

Instrumentation

In emission spectrography, the excitation of the sample is accomplished by thermal or electric means. Then the emitted radiation passes through a monochromator which allows only a particular wavelength and, finally, it enters the detecting system which measures the radiation. Different components of an emission spectrometer are as follows:

Sources

Different sources of excitation are given below:

- Flames: Molecules not requiring high temperatures can be excited by the use of flames. Flame temperature is dependent on various factors such as, type of fuel and oxidant, ratio of fuel to oxidant, burner type, the region in flame that is focussed into the entrance slit of the spectral isolation unit. Although flames are cheap but they are stable and reproducible sources having ability to handle many analytical issues.
- Direct current arc: This is one of the most widely used excitation sources for both quantitative and spectrochemical emission analysis.
- Advantages:
  - High sensitivity.

  - Excitation energy being mainly thermal is enough for excitation all metal elements.
  - By use of this arc, emission species produced are neutral atoms rather than ions.

Following difficulties may be encountered while operating this arc:

  - As the current increases a fall in resistance is seen.
  - Temperature is not same along the length of the arc.
  - Fluctuations in current are seen because of the changing arc length and temperature.

- Alternating current arc: Here a transformer maintains a high voltage of 2000 to 5000 V. This high voltage is helpful to make the arc jump the gap.
- Alternating current spark: In the primary circuit a voltage of 110 to 220 is maintained which is stepped up to 10,000 to 40,000 V by the step-up transformer.

Electrodes

Self-electrodes and graphite electrodes are the two types of electrodes employed in emission spectrography. If the substance whose investigation is to be done is a good conductor and can withstand high temperatures, the material itself is used as the electrode. Such types of electrode are called as self-electrodes. If the material to be analysed is not a good conductor and cannot withstand high temperatures, it is kept in a cavity of lower graphite electrode whereas the upper electrode is also made up of graphite which is brought to a point in a pencil sharpener.

Sample Holder

The sample holder introduces the sample into the electrical discharge. Sample holders are of two types:

- Solid Sample Holders: There are different kinds of samples that may be analysed by the use of emission spectroscopy. Various metals and alloys, wood, plastics, paint, polymers, soil, sea water can be easily determined through this method. Analysis of animal tissue, body fluids, plant materials, and soil extracts is also possible.
- Liquid Sample Holders: Analysis of liquid samples may be done directly by numerous methods. Two commonly used sample holders for liquids are shown in Figure.

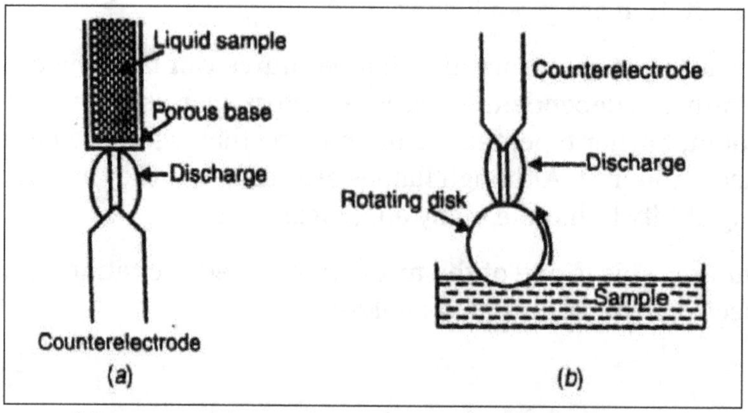

(a) Porous cup container and (b) Rotating disk electrode.

Monochromators

In emission spectroscopy both prism and grating monochromators are employed. A monochromator separates different lines of a sample's emission spectrum.

- Prism Monochromators: To be used as a monochromator a prism needs to be transparent to UV radiation. Quartz is used to make prisms. When light of a particular wavelength (say, emission line) passes through a prism, it emerges from the prism as two lines. Due to this half of the light's intensity is lost and thus the interpretation of the spectrum becomes difficult. These limitations can be overcome by employing two half prisms. This is also known as Cornu type prism.

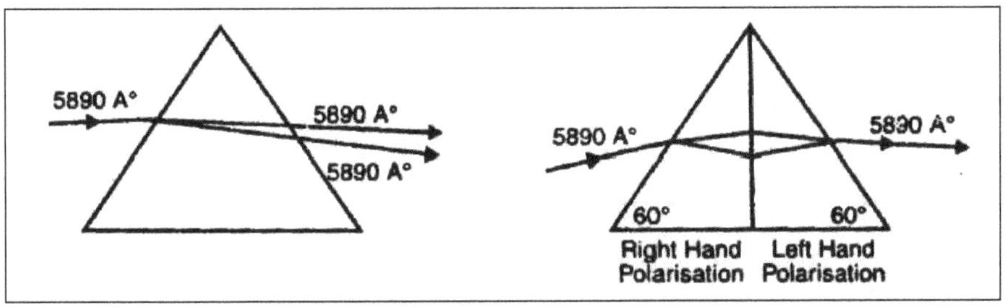

- Grating monochromators: Grating monochromators use in emission spectroscopy has made possible the development of sophisticated apparatus with high resolution. Even though the cost of a grating monochromator is high then also it has replaced prism as the dispersing element. As the grating gives linear dispersion, it is considered superior than a prism. The main drawback of grating monochromators is that higher-order wavelengths overlap but this can be overcome by use of filters or detectors which are not sensitive to higher orders.

- Echelle Monochromators: In recent times, Echellemonochromators have been widely used in emission spectrography.

Slits

All spectrographs are equipped with entrance and exit slits. Entrance slits keep out stray light and allow only the light from the sample to enter the optical path. Exit slits placed after the monochromator stop all but the desired wavelength range from entering the detecting system.

Detectors

Photo-multipliers and photographic plates or films are the two types of widely used detecting systems. Photographic plates or films are mainly employed for qualitative analytical purposes. For quantitative analysis, photomultipliers are used. Following are some of the advantages of using photographic plates over photo-multipliers for detection purpose in emission spectrography:

- Recording of a large number of spectral lines simultaneously is possible.

- Photographic plate provides a permanent record of the spectrum.

- The intensity of emission can be integrated by a photographic emulsion over a period of time.

- High sensitivity.
- Cheaper than an emission spectrograph equipped with photomultiplier detector.

The demerits of photographic plates for detection purpose are as follows:

- Controlled photographic development takes much time and elevates the risk of error.
- Photographic plates do not give quick response to spectral lines and their interpretation is very difficult.

Detection Through a Photographic Plate

The intensity of spectral lines is registered on a photographic emulsion in most of the spectrographs except the direct readers. This is developed by an appropriate developer in the given developing time at a definite temperature. Density of the film is then measured by a densitometer.

Detection by the use of Photomultipliers

Direct-reading emission spectrographs consist of photomultiplier tubes for detection instead of a photographic plate. These instruments are less popular than spectrographs consisting of photographic plates as a large number of photomultiplier tubes are required for detection of different emission lines simultaneously. Overall, detection through the use of photographic plates is employed for routine qualitative purpose whereas detection through photomultiplier is used for quantitative studies.

Spectrographs

Emission spectra may be analyzed by help of an optical arrangement that identifies component frequencies and their intensities. This optical arrangement is known by different names. For example, it is known as a spectroscope if it consists of a visual device; it is called as a spectrograph if recording of many wavelengths is done simultaneously either by photographic means; it is termed as a spectrometer if scanning of a spectrum is done. Except spectroscope, both spectrograph and spectrometer determine intensities. The different spectrographs may be differentiated primarily by whether they use a prism or grating as the dispersion medium.

Prism Instruments

The following figure shows the optical system of a simple prism spectrometer. The various components are an excitation source, an entrance slit, a collimator lens, a dispersing prism and an eye piece through which spectrum is seen. The eye piece can also be replaced by a camera so that the entire spectra can be photographed directly with ease.

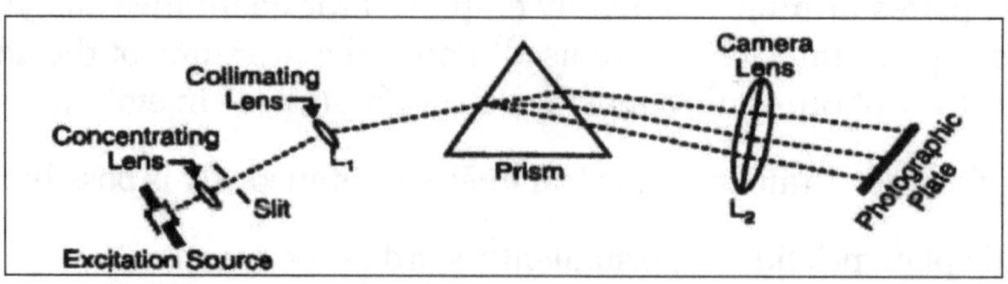

When an emission line passes through a quartz prism it undergoes splitting by polarisation and comes out from the prism as two lines. This limitation can be overcome by use of Cornu type prism which consists of two half prisms. Littrow type prism can also be used in place of Cornu type prism.

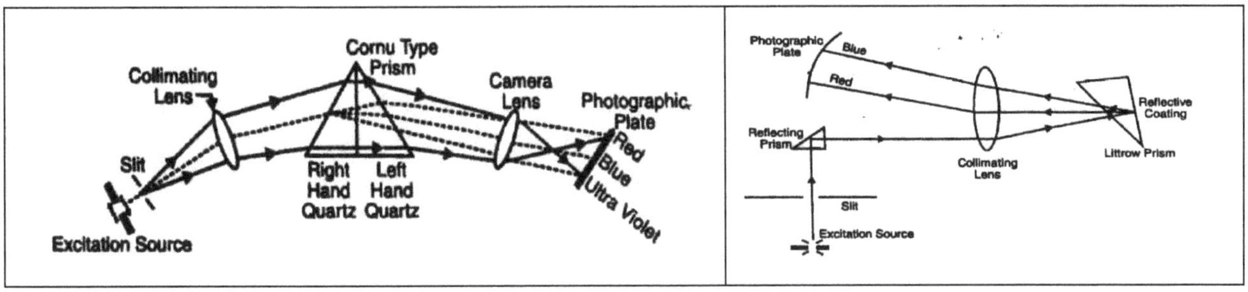

Grating Instrument

In principle, a grating can replace a prism in prism spectrograph keeping the rest of arrangement same. A grating spectrograph is chosen over prism because it gives linear dispersion and therefore the resolution is constant and is not dependent on wavelength. Commercial grating spectrographs are available with different mountings. One of these is Rowland mounting for a concave grating. In Wadsworth mounting, a concave mirror is employed along the concave grating. More compact equipment based on the Rowland circle is obtained by the use of Eagle arrangement. In Ebert mounting, instead of a concave grating a plane grating is employed.

Applications

Qualitative Determination

Emission spectroscopy is a destructive technique and offers the possibility of qualitative detection and quantitatively determination of most of the metals. With special adjustments, a few of the non-metals such as carbon, phosphorus and sulfur may also be analysed. It is for elemental determination purposes only and furnishes no information regarding the bonding in either molecules or alloys.

- Comparing Spectra and Lines: Qualitative detection of the elements may be accomplished by comparing the sample spectrum with the spectrum of pure known standard samples, or the wavelengths of the spectrum of the sample may be measured and then compared to values in standard tables.
- Last Lines, RU Lines: Raises ultima or RU lines are used for the qualitative detection of minor constituents and trace elements in a sample.

Quantitative Studies

Some of the methods used in quantitative spectrographic determination are:

- Comparison Sample Method: In this procedure, the spectra of the unknown sample and various standards are photographed alternately on the same plate under similar conditions. After development of the plate, the concentration of the desired component can be calculated by comparison of the blackening of the lines of this component with the same lines on the standards. Visual or photometric comparison of blackening may also be employed.

- Internal Standard Method: In this procedure, an internal standard is selected. Its known amount is then added to the sample. Both the sample (having internal standard) as well as internal standard emit spectral lines when they are excited through electrical discharge. These are photographed on a photographic plate.
- Spectrographic Standards: In addition to the choice of an internal standard, a most crucial phase in the development of a spectrographic procedure involves the preparation or acquisition of a set of standard samples from which a calibration curve is prepared.

Miscellaneous Applications

Emission spectroscopy has been used in various other areas to furnish useful information. Some of these are given below:

- The iron, steel and aluminium industries are based on emission spectroscopic studies in all steps of their processes. Nickel, chromium, silicon, manganese, molybdenum, copper, aluminium, arsenic, tin, cobalt, vanadium, lead etc., in iron and steel have been analysed routinely by use of an automatic emission spectrograph. Emission spectroscopy has also been employed for analysing different kinds of alloys.
- Oils: The catalysts can get poisoned by the trace elements present in erode oil used in the cracking process and therefore these elements need to be analysed on a regular basis to avoid such problems. Some of these trace elements are vanadium, copper; nickel and iron. All these have been determined by using an automatic emission spectrograph. Lubricating oils have also been examined during use to know the degree of wear of metal parts in motors and engines.
- Emission spectroscopy has been employed for the determination of various elements including sodium, potassium, zinc, copper, calcium, magnesium, nickel, and iron present in tissues of animals and men.
- Emission spectroscopy has been employed to determine various elements in plants and soils. This has helped in diagnosing various deficiency issues in plants.

Advantages

The merits of emission spectroscopy are as follows:

- An outstanding method for analysis of trace elements.
- Suitable for all metals and metalloids and sometimes for other elements as well.
- Analysis can also be done on a small sample.
- Time required for an analysis is short.
- The method can also be employed for routine quality control work and, if automated, can furnish analytical results on previously selected elements in a very short span of time. This is an ideal method for analysis of different samples like geological samples, metals and alloys, plants, forensic exhibits, soils, and environmental and biological samples.

The demerits of emission spectroscopy are as follows:

- An emission spectrometer is very costly if there is requirement of large instruments for analysis of heavy elements or very complex samples.
- The sample gets destroyed during analysis.
- This method is not suited for concentrated solutions.
- Use of emission spectroscopy is limited to the analysis of elements.

Flame Emission Spectroscopy

Flame photometry is based on measurement of intensity of the light emitted when a metal is introduced into a flame. The wavelength of the colour tells us what the element is, and the colours intensity tells us how much of the element is present. In early experiments, the visible colour of the flame was used to confirm the presence of certain elements in the sample, particularly alkali metals and alkaline-earth metals. Later the whole ultraviolet and visible range was utilized using a spectrophotometer. This instrument permitted us to select the wavelengths of the radiation and measure its intensity with considerable accuracy.

The spectrophotometric technique has proven to be one of the most reliable and easily used techniques for the determination of concentrations of sodium, potassium, calcium, and magnesium. Flame photometry is also named as flame emission spectroscopy because of the use of a flame to provide the energy of excitation to atoms introduced into the flame.

Flame photometry provides high sensitivity for determination of elements such as sodium, magnesium, strontium, potassium, lithium, calcium, and barium. Flame photometry is also used for determination of certain transition elements, such asiron, copper and manganese. Qualitative analysis by employing flame photometer is severely limited. Non-metal can also be analysed through this technique.

Limitations of Flame Photometry

The concentration of the metals may be measured by flame photometry, but frequently that technique is not the method of choice because of the relatively low energy available from a flame. Flame photometry tells little or nothing about molecular form of that metal in the original sample. It has not been used for the direct detection and determination of the noble metals, halides, or inert gases. All of these elements require more energy than the flame provides in order to become excited.

General Principles of Flame Photometry

Processes involved in flame photometry are given below:

- Solvent is vaporised, leaving particles of solid salt.

- Salt is vaporised or converted into gaseous state.
- A part or all of the gaseous molecules are dissociated to give free neutral atoms or radicals which are excited by thermal energy of the flame. The excited atoms emit photons and return to lower energy state. The measurement of the emitted photons, i.e., radiation, forms the basis of flame photometry.

The permitted energy levels of all the atoms can be represented in the form of diagram. Such diagrams are sometimes known as Grotrian charts.

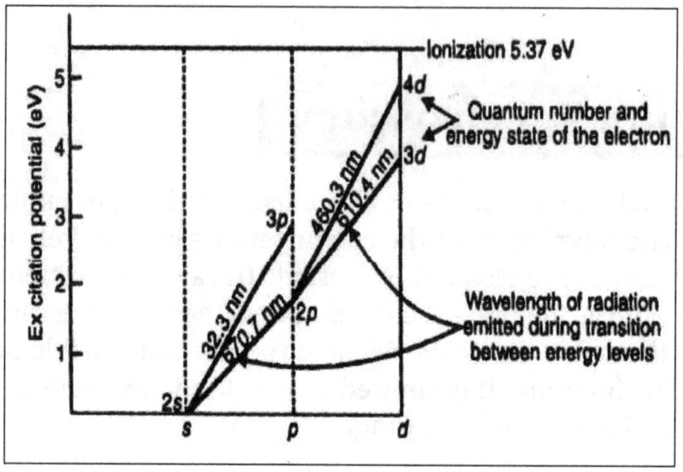

If E_2 and E_1 represent the energy of the higher and lower energy levels concerned, the radiation emitted during the jump may be defined by equation below:

$$E_2 - E_1 = hv$$

where h is the Planck's, constant and v the frequency of emitted light which is defined as follows equation below:

$$v = \frac{c}{\lambda}$$

On combining equations $E_2 - E_1 = hv$ and $v = \frac{c}{\lambda}$, we get equation:

$$E_2 - E_1 = \frac{hc}{\lambda} \lambda = \frac{hc}{E_2 - E_1}$$

From above equation one can calculate the wavelength of the emitted radiation which is characteristic of the atoms of the particular element from which it was emitted.

When flame photometry is employed as an analytical tool, the wavelength of the radiation coming from a flame tells us about the elements which are present in that flame. Also, the intensity of the radiation enables us to know the amounts of those elements present. The following is a concise description of the sequence of events that normally occurs in flame photometry. A portion of the neutral atoms or radicals formed in the flame may combine to form new gaseous compounds.

As the formation of these compounds decreases the population of neutral atoms and hence the intensity is also decreased, this constitutes the chemical interference in flame photometry.

$$\frac{N^*}{N_o} = \Delta e^{-\Delta E/kT}$$

where N° is the number of excited atoms, No the numbers of atoms remaining in the ground state, A a constant for a particular element, ΔE the difference in energies of the two levels, k the Boltzmann's constant, and T the temperature of flame.

The fraction of free atoms that are thermally excited is governed by a Boltzmann distribution which is as follows:

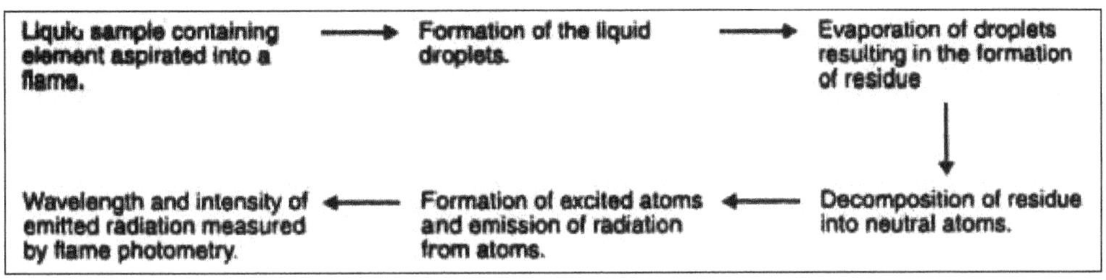

The temperature of fuel is controlled by the type of fuel and oxidant used. Some flame temperatures are given in Table:

Fuel	Oxidant	Flame Temp. (°C)
H_2	O_2	2800
H_2	Air	2100
H_2	Ar	1600
Acetylene	O_2	3000
Acetylene	Air	2200
Acetylene	N_2O	3000
Propane	O_2	2800
Propane	Air	1900

The energy available in the flame for exciting atoms is dependent on the flame temperature as defined by the Boltzmann distribution. Flames have only limited amounts of energy available when compared to excitation sources such as emission spectrographs or plasma emission torches (ICP). For this reason, flames are most useful only for elements that require low amounts of energy to become excited. They are not the method of choice for transition metals and most of the other metals in the periodic table.

These metals generally require significantly more energy in order to become excited. They are most often done by emission spectrography or plasma emission where much higher energy is available. However, it must also be stated that when we try to analyze group I and group II elements in these high energy sources, we run the risk not only of exciting them but of ionizing them. Although the Boltzmann distribution would lead us to believe that the emission intensity would be greatly increased in plasma emission for the group I and group II elements, in practice it is found not to be so because the atoms are ionized and this results in a loss of atoms and the generation of ions.

Ionization causes a complete change of energy levels and therefore the entire emission spectrum, with a reduction in the intensity of the atomic spectra. The flame is a source of intense radiation especially for atoms inside the flame. It is probable that, this is a source of excitation of the atoms (radiation excitation) which augments the thermal excitation, expected from the Boltzmann distribution.

This point has not been extensively studied but provided a catalyst for the evolution of atomic fluorescence. For example, a fuel-rich oxyacetylene flame generates intense radiation bands at short, wavelengths (300-200 nm). Iron solutions injected into these flames strongly emit iron lines at many wavelengths down to 200 nm. On the other hand, oxygen-rich flames operating at the same temperature do not emit such broad-band radiation, and iron lines are not emitted from such flames.

The emission intensity generated by thermally excited atoms should be the same according to the Boltzmann distribution but is clearly not the same. The advent of the nitrous oxide-acetylene flame has provided us with a flame with higher temperature than was hitherto available. This has increased the usefulness of flame photometry. Also, the use of the elongated burner as developed for atomic absorption spectroscopy has increased the sensitivity of the method.

Because flames are a low-energy source they populate the first excited energy state E_1 reasonably well but the upper energy states E_2 E_3 etc. only poorly. Flame photometry therefore utilizes the resonance transitions $E_1 \rightarrow E_0$ most commonly and other transitions only under special circumstances.

Instrumentation

Equipment used in flame photometry is shown in Figure. The instrument possesses the same basic components as a spectroscopic apparatus has. However, the basic components are source, monochromator, slit system and detector system. The flame photometer also includes a burner which is utilised for burning the sample solution and exciting the atoms produced in the flame after burning.

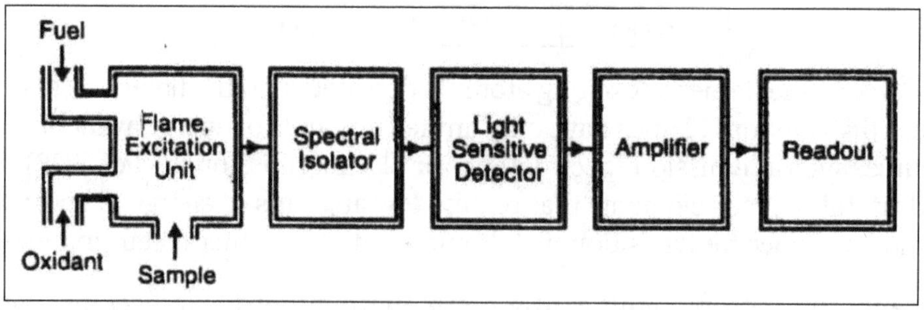

Block diagram of a flame photometer.

Burner

The flame used in the flame photometer must possess the following functions:

- The flame should possess the ability to evaporate the liquid droplets from the sample solution, resulting in the formation of solid residue.

- The flame should decompose the compounds in the solid residue formed in step (i), resulting in the formation of atoms.
- The flame must have the capability to excite the atoms formed in step (ii) and cause them to emit radiant energy. For analytical purposes, it becomes essential that emission intensity should be steady over reasonable periods of time (1-2 min).

Flame temperature is controlled by several factors which are as follows:

- Type of fuel and oxidant and fuel-to-oxidant ratio.
- Type of solvent for preparing sample solution.
- Amount of solvent entering into flame.
- Type of burner employed.
- The particular region in flame which is to be focussed into entrance slit.

In flame photometry, several burners and fuel-oxidant combinations have been used to produce the analytical flame. Some of these are discussed as below:

Meeker Burner

This burner was used earlier and employed natural gas and oxygen. As this burner produced relatively low temperatures and low excitation energies, this was generally used for the study of alkali metals only. The flame produced by Meeker burner is not homogeneous chemically. It means that there are different regions in the flame, i.e., an "oxidising" region and a "reducing" region appear in the flame.

Total Consumption Burner

A typical total consumption burner is shown in Figure. In this burner, the fuel and oxidant are hydrogen and oxygen gases respectively. In this type of burner, the liquid sample is drawn into the flame. From the side tubings, hydrogen and oxygen are entering and both are burning at the top of the burner to produce a flame. As soon as the liquid sample is drawn into the base of flame, the oxygen aspirates sample solution leaving a solid residue. Atomisation and excitation of the sample then follow:

- The name "total consumption burner" is used because the entire sample that enters the capillary tube will enter the flame regardless of droplet size.
- The flame produced by total consumption burner is noisy and turbulent but can be adjusted to produce high temperatures by proper control of fuel-to-oxidant ratio.
- This burner was used for a number of years but its use has been stopped and replaced by other types of burners.
- As the processes leading to atomic excitation in the flame differ in the oxidising and reducing regions in flame, different concentrations of excited atoms are obtained in these regions.

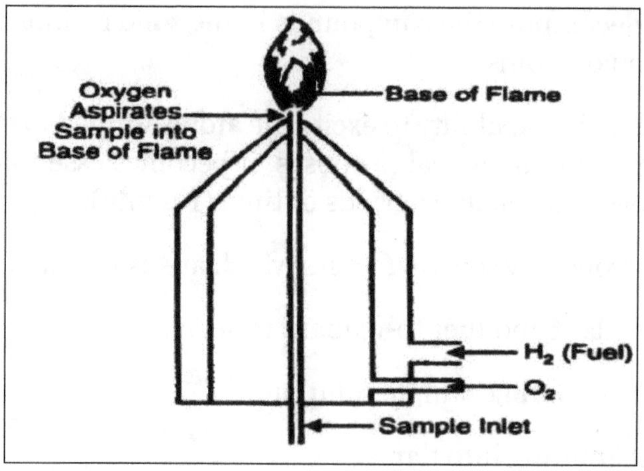

Premix of Laminar-Flow Burner

In this type of burner, aspirated sample, fuel and oxidant are mixed before reaching the burner opening. An important feature of this burner is that only a small portion (about 5%) of the sample in the form of small droplets reaches the flame and is easily decomposed. By the easy decomposition in means that an efficient atomization of the sample in the flame will take place.

Larger droplets from the aspirator impinge on the side of the spray chamber and are drained off. Thus, in this burner 95% of the sample may be wasted, thereby, resulting in a loss of sensitivity. However, this loss must be balanced against the loss of larger droplets.

The flame produced by premix burner is non-turbulent, noiseless and stable. Another advantage is that in this burner easy decomposition of the sample takes place which results in an efficient atomization of the sample in the flame. Premix burners can handle solutions up to several percent without clogging. The main disadvantage of premix burner arises when the sample contains two solvents.

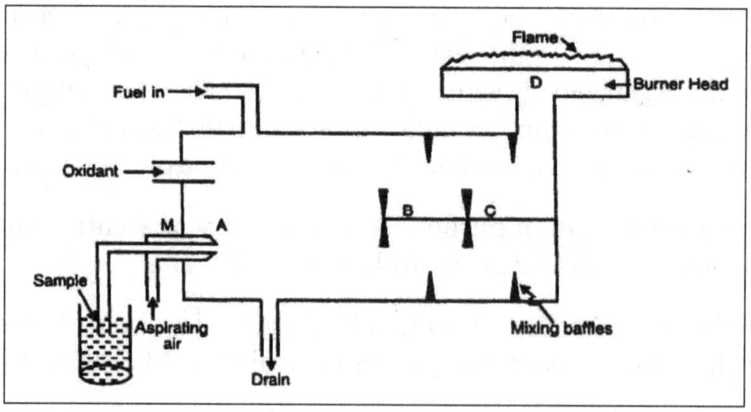

Lundegardh Burner

In the Lundegardh burner, the sample must be in liquid form. It is aspirated into the spray chamber. Large droplets condense on the side and drain away; small droplets and vaporised sample are swept into the base of the flame in the form of a cloud. Various devices have been used to enhance the nebulization stage in this type of burner. These include the use of the impact bead, ultrasonic vibrators, and more recently thermospray heaters. An important feature of this burner is that only about 5% of the sample reaches the flame. The rest of the droplets condense and are drained away. This is a significant loss in atomization efficiency and, therefore, sensitivity.

Difficulties may also arise if there is any selective evaporation of the solvent in the spray chamber. In particular, if there are two solvents, the more volatile will preferentially evaporate, leaving the element of interest in the less volatile component. The latter may drain away, taking the sample with it. In this event many of the sample atoms never reach the flame, and the emission intensity is reduced and an incorrect analysis obtained.

The atomization step is very dependent on the efficiency of the nebulization since the droplets must be evaporated and the residue broken down to liberate free atoms after introduction into the flame. A number of factors affect the drop size d_0 including viscosity η, density ρ surface tension γ of the sample solution, gas and the quantity of aspirated solution, ϕaq liquid, and the velocity of the nebulizing gas. These variables are related in the empirical expression.

$$d_0 = \frac{585}{v}\left(\frac{\gamma}{\rho}\right)^{0.5} + 597\left(\frac{\eta}{(\gamma\rho)^{0.5}}\right)^{0.45} 1000\left(\frac{\phi \text{liq}}{\phi \text{gas}}\right)^{1.5}$$

The diameter d_0 controls the volume-to-surface relationship. It can be seen from expression that it is important to maintain constant viscosity, surface tension, density and flow rates of the pertinent gases to reproduce the nebulized drop size and therefore the emission atomization process intensity. To that end it is essential that the standards and sample be analyzed under identical flame conditions and that the solvents and matrices are as close to each other as possible.

Table: Results Obtained with Shielded Burner.

Element	Sensitivity (ppm)
Ba	0.05
Bi	2
Ca	0.002
Cr	0.0007
Co	0.04
Pb	0.05
Mg	0.3
V	10

Shielded Burners

T. S. West al. developed shielded burners in which flame is shielded from ambient atmosphere by a stream of inert gas. This shielding leads to better analytical sensitivity. Table shows results obtained with commercial equipment based on this technique and developed by the Beckman Instrument Co.

Nitrous Oxide-Acetylene Flames

During research in atomic absorption spectroscopy, J, Willis and M. Amos independently found that nitrous oxide-acetylene flames were superior to other flames for efficiently producing free atoms. This was particularly true for metals with very reflective oxides, such as aluminium and titanium. Later, workers in the field of flame photometry found that this same flame was very useful in flame photometry. However, the high temperature reduces its usefulness for the determination of the alkali metals because they are easily ionised.

One problem encountered with this type of flame was the intense background emission, which makes measurement of the metal emission very difficult. However, the "wobbler" designed by Rains for background correction and the commercial nitrous oxide-acetylene burners developed provide equipment capable of high sensitivity and accuracy. Note the linear relationship between the emission intensity and the concentration of the metal. Note also the extended useful analytical range to the method. The calibration data are drawn to log-log paper for convenience.

Table: Percent Ionisation in Flames.

Element	Ionization potential (eV)	Acetylene–air (2400 °C)	Acetylene–oxygen (3100 °C)	Acetylene –Nitrious oxide (3200 °C)
Lithium	5.39	0.01	16	–
Sodium	5.14	1.1	26	–
Potassium	4.34	9.0	82	–
Rubidium	4.18	14.0	89	–
Cesium	3.48	29.0	96	–
Magnesium	7.64		0.01	6
Calcium	6.11	0.01	7	43
Strontium	5.69	0.01	17	84
Barium	5.21	2	42	88
Manganese	7.43			5

Sequence of Events in a Flame

The sequence of events occurring in the flame are given below:

- Solvent is evaporated, leaving minute particles of salt.
- Salt is vaporized or converted into gaseous state.
- A part or all of the gaseous molecules are dissociated to give free neutral atoms or radicals.
- A part of neutral atoms may be thermally excited or even ionized.
- A portion of neutral atoms in the flame may combine to form new gaseous compounds.

Mirrors

Radiation from flame is emitted in all directions. Much of the radiation is lost. In order to maximize the amount of radiation used in the analysis, a mirror is located behind the burner to reflect the radiation back to the entrance slit of the monochromator.

The reflecting surface of the mirror is front-faced. If the reflecting surface were on the rear as in the normal household mirror, the radiation from the flame would have to go through the support material such as glass or quartz twice before it is reflected to the entrance slit. Since the support material absorbs some radiation, there would be a considerable loss of signal, particularly at the shorter wavelengths. Front surface mirrors are most efficient, but they are not physically protected. Care should be taken to protect them by keeping the instrument away from corrosive atmospheres.

Slits

With the best equipment, entrance and exit slits are used before and after the dispersion elements. The entrance slit cuts out most of the radiation from the surroundings and allows only the radiation from the flame and the mirrored reflection of the flame to enter the optical system. The exit slit is placed after the monochromator and allows only a selected wavelength range to pass through to the detector.

For many purposes it is essential that this wavelength range be very narrow, that is, of the order of a few nanometers. This is necessary if emission lines from other components in the flame have a wavelength similar to those of the emission lines of the elements being determined. The slit must prevent such interfering lines from reaching the detector.

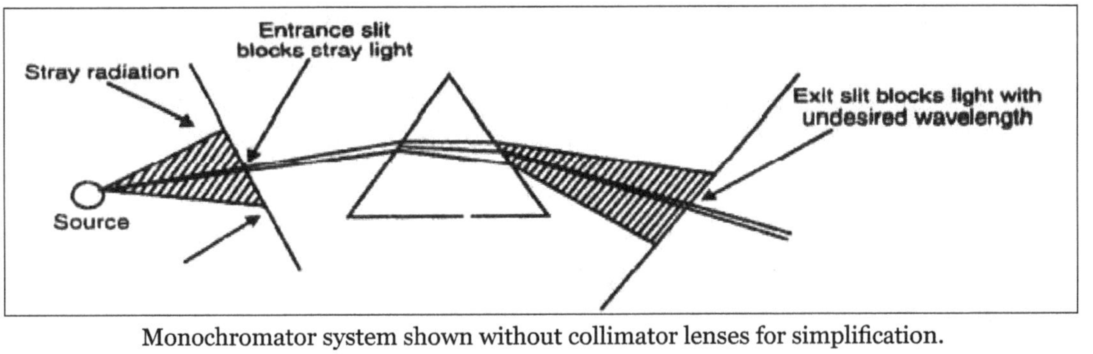

Monochromator system shown without collimator lenses for simplification.

Monochromators

In simple flame photometers, the monochromator is the prism. But in expensive models, the grating monochromators are used. Quartz is the material most commonly used for making prisms even though its dispersive power is less than that of glass as it is transparent over the entire region. The grating monochromator employs a grating which is a series of parallel straight lines cut into a plane surface.

Filters

In some elements, the emission spectrum contains a few lines. In such cases wide wavelength ranges will be allowed to enter the detector without causing any serious error. In such a situation an optical filter may be used in place of the slit and monochromator system. The filter is made from such a material which is transparent over a narrow spectral range. When a filter is kept between the flame and detector, the radiation of the desired wavelength from the flame will be entering the detector and be measured. The remaining undesired wavelength will be absorbed by the filter and not measured.

The flame photometers which use filter monochromators are very convenient for simple repetitive analysis. However, such instruments can be used for a small number of elements. The reason for this is that a large number of filters are employed.

Detectors

Detector measures the intensity of radiation. It should be sensitive to radiation of all wavelengths that may be examined. In good flame photometers, the photomultiplier detectors are employed which produce an electrical signal from the radiation falling on them.

Effect of Solvent in Flame Photometry

The solvent used in the sample affects the signal in two ways. First, viscosity controls the rate at which the sample is aspirated into the flame. There is an optimum sample flow rate, which has to be experimentally determined. If the flow rate is too great, the flame is swamped and the signal drops off; if the flow rate is too low, then the signal is decreased because insufficient sample finds its way into the flame.

The second effect of the solvent is caused by the difference between aqueous or organic materials. If it is aqueous, then the sample requires energy to evaporate it. Generally an inorganic salt is left, which requires more energy from the flame to decompose it. These are two endothermic steps, which slows down the atomization process. On the other hand, if the solvent is organic, it burns on introduction to the flame and usually leaves an organic residue, which in turn burns inside the flame.

Each of these steps is exothermic, the atomization efficiency is increased, and there is an enhancement of signal. This is the reason for signal enhancement when organic solvents are used rather than aqueous solvents. It can be seen from these comments that in order to get good quantitative data it is necessary that the sample introduction note, the solvent used, and flame conditions be the same for both the samples to be analyzed and the preparation of calibration curves. Any variations in these conditions cause a major variation in flame intensity and therefore in the analytical data obtained.

Instruments

There are a large varieties of flame photometers in widespread use, ranging all the way from very single-beam filters photometers to multichannel spectrometers with automatic background correction.

First Type

The simplified diagram of a flame photometer is depicted in Figure. The burner used in flame photometer shown in Figure is a total consumption type, but a premix burner can also be used. In Figure, a collimating mirror is used behind the flame to increase the emission intensity. But this mirror is often omitted when premix burners having a long flame path are used. This is sometimes, necessary to remove the backing mirror in multiple-beam instruments where a sufficient room is not available to include such mirrors.

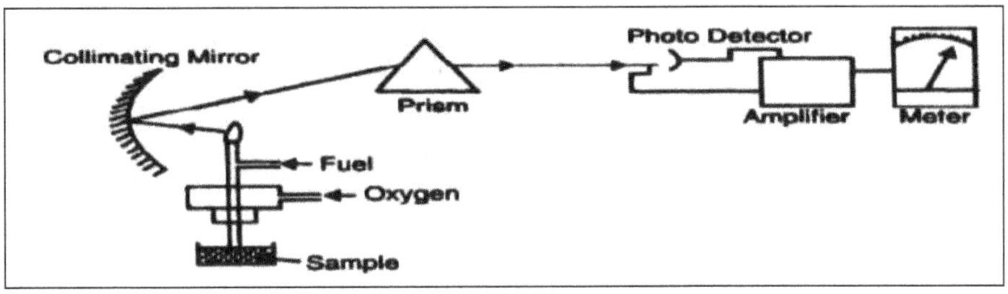

The sample solution is sucked by an atomiser operated by one of the flame producing gases. Then the sample solution is aspirated into the flame in the form of a fine spray. Spectral emission comes from the excited atoms formed during the process of combustion in a flame.

The emitted radiation is collected by a collimating concave mirror from the flame and is then permitted to pass through a prism and slit. The radiation of the appropriately selected wavelength strikes a photo detector and the magnitude of the electrical signal developed is read out on a meter.

Second Type

Another type of flame photometer is internal standard flame photometer. In Figure an internal standard flame photometer is illustrated. Lithium is used as an internal standard and an equal concentration is added to the standard and sample solutions. The sample solution containing internal standard (lithium) is sucked by an atomiser operated by one of the flame producing gases and as a fine spray is fed into the flame. The emitted radiation is collected by a mirror through a filter.

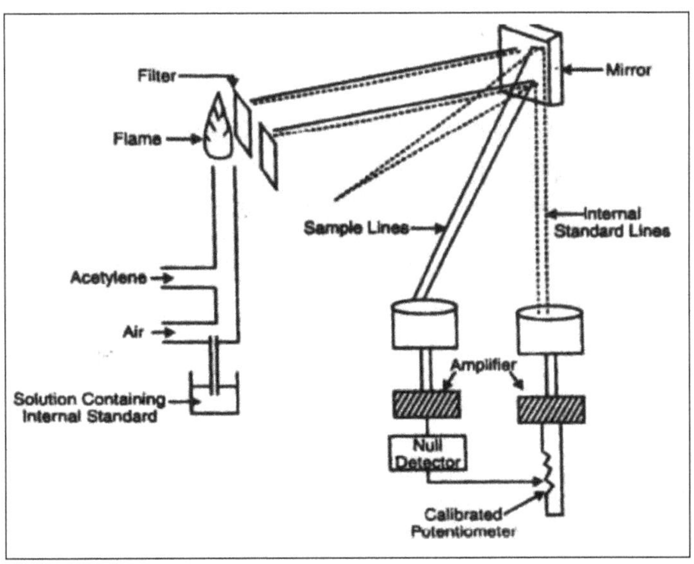

Schematic diagram of an internal standard photometer.

The emitted radiation from the mirror is divided into two parts. One part arises due to the internal standard (lithium) whereas the second part arises due to the presence of element to be measured. Both these parts are received by separate amplifiers and are then fed to the common detecting system which records the intensity of the element (to be determined) to that of internal (lithium). Thus, this flame photometer gives a direct and simultaneous reading of the ratio of intensities.

The use of this flame photometer removes the effects of momentary fluctuations in flame characteristics caused by fluctuations in fuel or oxidant pressures. Therefore, internal standard flame photometer is better than the first type. In internal standard flame photometer, the errors due to differences in viscosity and surface tension are minimised considerably. However, the effect of radiation mayor may not be reduced, depending On the similarity of the effects on the two elements.

Applications of Flame Photometry

The Qualitative Analysis

Flame photometry is only used to detect elements of groups I and II. These elements are lithium, potassium, sodium, magnesium, calcium, strontium, and barium. This method has some limitations which are as follows:

- This method is less reliable than atomic absorption spectroscopy.
- This does not provide information about the molecular structure of compound present in the sample solution.
- Non-radiating elements, such as carbon, hydrogen and halides cannot be detected by flame photometer. These can only be determined under special circumstances. For example, if chlorine is to be detected in a liquid sample, the best method is to precipitate it as silver chloride which is then aspirated into flame of a flame photometry to carry out the determination of silver. From the result, the chloride content can be calculated.

Quantitative Analysis

This is one of the most useful applications of flame photometry. This is used for rapid quantitative determination of the elements in groups I and II of the periodic table. If high optical resolution equipment is used, other metallic elements besides that of I and II groups can also be determined.

In the Table, there is a list of elements with their emission wavelengths and detection limits. All the elements present in this table have been determined successfully by flame photometry.

Element	λ Å	Detection limit ppm	Element	λ Å	Detection limit ppm
Aluminium	3962 4840	0.5 0.5	Copper	3247	0.6
Antimony	2528	1.0	Iron	3720 3860	2.5 2.7
Arsenic	2350	2.2	Lead	4058	14
Barium	4555 5536	3 1	Lithium	6708	0.067

Bismuth	2231	6.4	Magnesium	2852 3830	1.0 1.6
Calcium	4227 5540	0.07 0.16	Mercury Nickel	2536 3524	2.5 1.6

Experimental Procedure for Quantitative Analysis

Whenever quantitative analysis is to be carried out, the first job is to introduce the sample solution into flame and then intensity of radiation is measured at pertinent wavelength. From the radiation intensity, one can determine concentration of metal present in the sample by employing one of the following methods: Standard addition method and Internal standard method.

Standard Addition Method

In this method, the various solutions containing the unknown and known amounts of standard substance are used. In this method, the first job is to know the signal intensity for the unknown X by aspirating its solution into the flame of a flame photometer. Then a series of solutions having the unknown X plus varying amounts of the standard A are prepared and their signals obtained by aspirating them into the flame.

A plot is drawn between the signal intensity (Y-axis) against concentration of X plus concentration of standard A. From the plot, the concentration of the unknown can be determined from the intersection of the curve with the concentration axis. A typical curve illustrating this method is shown in figure:

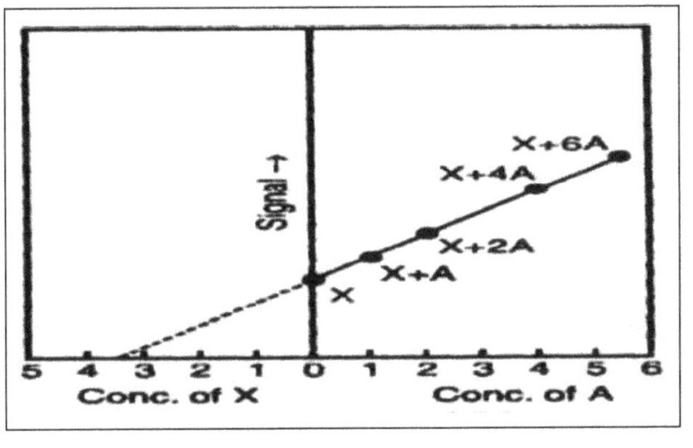

Advantages

- The method is useful for determining very low concentrations of the elements present in the sample.
- This method compensates for any unexpected interfering material present in the sample solution.
- This method can be used to estimate such elements which are rarely analysed.

Precaution

Whenever standard addition method is to be employed, a correction should be made for background emission from the flame. This can be done by injecting the solvent into the flame and then its signal intensity, called background emission, is measured. When calculations are to be made, the intensity of background emission must be subtracted from the emission due to sample.

Internal Standard Method

When this method is applied to a simple flame photometer, following steps are involved:

- Prepare standard sample solutions containing known amounts of internal standard element. Then aspirate these standard samples into the flame and record the signal intensity for each of these samples.
- Now aspirate the samples having unknown concentrations into the flame and record the signal intensity.
- The ratio of the two intensities is calculated and is then plotted vs. the concentration of the element to be determined.

A calibration curve is obtained. From this, one can calculate concentration of the element present in the sample if its signal intensity is known by aspirating it into the flame. When this method is used with a simple flame photometer, the effects of momentary fluctuations in flame characteristics will not be eliminated. However, these fluctuations can be overcome by using a direct-reading instrument which gives a direct and simultaneous reading of the ratio of two intensities. Every element cannot be chosen as the internal standard. However, it should fulfil the following characteristics:

- Its amount should be very small in the original sample.
- It should give an emission line which reacts to interferences as the line of the does.

One or both of these requirements are often violated, sometimes with disastrous results.

Miscellaneous Applications

Flame photometry is very helpful in analysing biological fluids and tissues. Flame photometry has been used for determining sodium, potassium, aluminium, calcium, cobalt and iron in soil analysis. It has also been used for petroleum products, natural and industrial waters, glass, cement, and metallurgical products. Boron has been determined in various types of organic compounds.

Interferences in Flame Photometry

The radiation intensity may not accurately represent the sample concentration because of the presence of other materials in the sample. These materials result in interference in the analytical procedure. It is only through adequate control of this interference that flame photometry would provide good analytical results. Following are the more commonly encountered interference processes in flame photometry:

Spectral Interference

This type of spectral interference arises when two elements or compounds possess different spectra but their spectra may partly overlap and both are emitting at some particular wavelength. This type of interference is more common at high flame temperatures. An interesting example is that the iron line at 3247.28 A° overlaps the copper line at 3247.54 A° and the iron line at 2852.13 A° overlaps the magnesium line at 2852.12 A°.

This type of error can be overcome by removing the effect of interfering element effect by using extraction method or using calibration curves prepared from solution having similar quantities of interfering element. The second type of spectral interference can occur if spectral lines of two or more elements are close but their spectra do not overlap. The possibilities of such interference can be decreased to a considerable extent by enhancing resolution of spectral isolation system.

A third type of spectral interference occurs between a spectral line and a continuous background. This type of interference is encountered due to high concentration of salts in the sample. This occurs especially in salts of the alkali and alkaline earth metals. A continuous background is also produced by some organic solvents. An example is methyl isobutyl ketone which is particularly troublesome in flame photometry. This type of interference can be corrected by scanning technique.

Ionisation Interferences

In some cases, some of the metal atoms may ionise in high-temperature flame, e.g:

$$Na \rightleftharpoons Na^{+} + e^{-}$$

Sodium ion possesses an emission spectrum of its own, with different frequencies from those of the atomic spectrum of sodium. Thus, ionisation decreases the radiant power of atomic emission.

The interference due to ionisation can be overcome by addition of a large quantity of a potassium salt to all of the solutions-unknown and standards. The addition of potassium prevents the ionisation of sodium but it itself undergoes ionisation. Thus, the sodium atom emission is enhanced. This type of interference is restricted to the elements of the first group of the periodic table.

Cation-Cation Interference

Cation-cation interferences decrease signal intensity of the element present. An example of cation-cation interferences is that aluminium interferes with magnesium and calcium. Also, potassium and sodium have cation-cation interference on one another.

Oxide Formation Interference

This type of interference occurs due to formation of stable oxides. Thus, emission intensity is lowered. All of the alkaline earth elements form oxides and are subject to this type of interference.

Cation-Anion Interference

Presence of certain anions in solution may affect intensity of radiation emitted by an element. For example, anions such as phosphate, oxalate, aluminate and sulphate may affect intensity of radiation emitted by an element. For example, calcium in presence of phosphate ion forms a stable substance, so the calcium signal is depressed because it will not decompose so easily, resulting in production of lesser atoms. Another example is that a given concentration of barium sulphate produces low emission intensity than the same concentration of barium chloride, because barium chloride can be broken down more easily than barium sulphate.

This type of interference can be removed either by extraction of the anion or by using calibration curves which are prepared with the same predominant anion at the same concentration in the standard solution as that found in the sample. This type of interference can be overcome by either using very high temperature flames to dissociate those oxides producing free atoms for excitation or using oxygen-deficient environment to produce excited atoms.

Numerous factors will influence intensity of light emission from a given solution. Some of these are as follows:

- Viscosity: The addition of a substance which increases the viscosity of the solution decreases the intensity of light emission. This decrease results in due to a reduction in the efficiency of atomization.
- Presence of Acids: When an acid is present in the sample solution, this decreases the light intensity. This decrease arises due to the disturbance of the initial dissociation equilibrium.
- Presence of Other Metals: If other metals are present, these also alter the intensity of emitted radiation. In order to remove this defect, special filters are used which will absorb radiation due to the element which is to be estimated in the sample solution.

Determination of Non-Metals

The determination of non-metals by flame photometry has two major handicaps. First, the atomic state is usually not stable. For example, chlorine does not exist as Cl but as Cl_2. Second, the resonance lines are in the vacuum ultraviolet. High energy is necessary to cause electronic excitation-more energy than is available in conventional flames. Atomic emission of non-metals has not been exploited in flame photometry until recently and even then only using research grade equipment. However, molecular emission has been observed. These are band spectra rather than line spectra and are therefore less intense, analytically less sensitive, and more open to spectral interferences.

Raman Spectroscopy

There are numerous forms of light-matter interaction: fluorescence and phosphorescence are examples of absorption and subsequent emission of light by matter. Elastic scattering of light, such as Rayleigh scattering by atoms, molecules or phonons, and Mie/Tyndall scattering by dust particles are examples where the wavelength of the light is unchanged. Inelastic scattering such as Brillouin scattering by acoustic waves in crystals, Compton scattering by charged particles and Raman scattering by molecules or phonons are examples where the wavelength of the light does change. Raman scattering of light by molecules was first predicted using classical quantum theory by Smekal in 1923 and experimentally observed by Raman and Krishnan in 1928.

There are now more than 25 different types of known Raman spectroscopy techniques, such as spontaneous Raman, hyper-Raman scattering, Fourier transform Raman scattering, Raman-induced Kerr effect spectroscopy and stimulated/coherent Raman scattering. Fifty years after its first observation, Raman spectroscopy started to become a prominent analysis technique among other optical metrology techniques, such as those involving absorption of infrared light; particularly when water and other useful polar solvents were present, because these media typically strongly absorb light in the infrared region. For example, in 1974, Fleischmann et al. used Raman spectroscopy to distinguish two types of adsorbed pyridine (a basic cyclic heterodyne compound molecule) on the surface of a silver electrode to mitigate absorption effects. This experiment was incidentally the first serendipitous observation of SERS.

Raman spectroscopy is now an eminent technique for the characterisation of 2D materials (e.g. graphene and transition metal dichalcogenides) and phonon modes in crystals. Properties such as number of monolayers, inter-layer breathing and shear modes, in-plane anisotropy, doping, disorder, thermal conductivity, strain and phonon modes can be extracted using Raman spectroscopy.

The biological and medical fields of research are greatly impacted by the development of Raman spectroscopy as it is a label-free (does not require fluorescent marker molecules) chemically selective hyperspectral imaging technique. For instance, studying the transdermal delivery of drugs into skin often ordains ex vivo and invasive analysis techniques. Ex vivo transdermal delivery studies are unfavourable because skin regeneration stops, the immune response ceases, and metabolic activity is usually lost. Hence, the performance of transdermal drug delivery ex vivo is not an accurate reflection of the in vivo situation. However, non-invasive in vivo measurements can be performed using Raman spectroscopy to gain detailed information about the molecular composition and concentration gradients in the skin. In many biological processes, living microorganisms such as bacteria act as biocatalysts. Raman spectroscopy can probe inhomogeneity in the properties and physiological status of individual cells in biocatalytic processes. Raman spectroscopy has also been used to identify and differentiate benign and malignant breast cancer lesions by probing their unique chemical compositions.

For biological samples, approximately 90% of the peaks are found in the 'fingerprint' spectral region, covering ($\Delta\tilde{\nu} \sim 500\,cm^{-1}$ to $\sim 1800\,cm^{-1}$; $\Delta\tilde{\nu}$ is the wavenumber shift), with the remaining found in the higher energy CH/OH stretching vibrational modes covering ($\Delta\tilde{\nu} \sim 2700\,cm^{-1}$ to $\sim 3300\,cm^{-1}$).

Coherent Raman Spectroscopy

Coherent light-scattering events involving multiple incident photons simultaneously interacting with the scattering material was not observed until laser sources became available in the 1960s, despite predictions being made as early as the 1930s. The first laser-based Raman scattering experiment was demonstrated in 1961.

SRS is a coherent process providing much stronger signals relative to spontaneous Raman spectroscopy as well as the ability to time-resolve the vibrational motions. SRS is relevant to numerous areas of research such as plasma physics, atomic interferometry, supercontinuum generation, imaging of biomolecules in food products, imaging chemistry inside living cells, bulk media and nanoscale specimens. The exchange of photon orbital angular momentum by SRS in plasma is gaining interest, particularly in the context of inertial fusion research. Supercontinuum generation is a complex nonlinear phenomenon that is characterized by the dramatic spectral broadening of intense light pulses passing through a nonlinear material. Knight et al. demonstrated flat ultrabroadband octave-spanning white-light supercontinuum generation by SRS and parametric four-wave mixing with 60-ps pump pulses of sub-kilowatt peak power in a photonic crystal fibre. Kasevich and Chu demonstrated a matter-wave interferometer with laser-cooled sodium atoms using the mechanical effects of stimulated Raman transitions. SRS has even been used to observe time-resolved vibrational spectra of the primary isomerisation of retinal in the visual pigment rhodopsin.

Since its resurgence in 1999, CARS has become a prominent vibrational mode imaging tool in biological medicine. As anti-Stokes photons are blue shifted from the pump and Stokes frequencies, they are more easily detected in the presence of single-photon fluorescence. CARS microscopy has been successfully applied to live-cell imaging, skeletal stem cells, tracing toxic nanomaterials in biological tissues, volumetric imaging of human somatic cell division, flow cytometry, detection of brain tumours and tracking organelle transport in living cells. Zirak et al. has developed a CARS endoscope for in vivo imaging and demonstrated the instrument with murine adipose tissue and human nervus suralis samples. Evans et al. have combined CARS with video rate microscopy to chemically image tissue in vivo. Potma and Xie have directly visualised lipid phase segregation in single lipid bilayers with CARS. CARS can even be used as a high temporal and spatial resolution thermography technique and has found applications in electronic and opto-electronic device characterisation and even turbomachinery.

Orientational order is a salient feature of many soft matter systems. Detail in structural molecular organisation is a prevailing goal in the field of biology, biomedicine, material sciences and molecular physics. Polarisation-resolved optical microscopy is becoming a powerful tool to address molecular orientational distributions into the focal volume of a microscope. In coherent nonlinear optics, polarised second harmonic generation, polarised third harmonic generation and polarised four-wave mixing have already been used to recover orientational information on endogeneous proteins and lipids in biological tissues. In addition to the orientational information, coherent Raman scattering (CRS) processes are sensitive to molecular bond vibrations, allowing chemical specificity without the need for fluorescence labelling/dyes. CARS microscopy can be used to image chemical and orientational order of liquid crystalline (commonly used in display technology) samples. Polarisation-resolved hyperspectral SRS microscopy has also been demonstrated as a label-free biomolecular imaging technique with teeth. In addition, polarised-CARS has been used to study the molecular order of lipids in myelin at sub-diffraction scales in mice.

Enhanced Raman Spectroscopy

The sensitivity of Raman spectroscopy can be enhanced through various techniques such as resonance Raman spectroscopy, TERS or SERS. SERS is particularly interesting since it allows an enhancement of several orders of magnitude of the Raman signal by modifying the surface upon which an analyte material is to be placed. The enhanced light-matter interaction in TERS and SERS is tuneable (to some extent) by modifying thc surfacc nanostructure of metallic films on dielectric surfaces. The wavelength of charge density oscillations, known as plasmons, is dependent on these surface nanostructures and can enhance the light-matter interaction locally.

Fleischmann et al. first observed SERS in 1974 when investigating pyridine on the rough surface of a silver electrode. The sensitivity of SERS makes it well-suited to study electron transfer reactions, which lie at the heart of numerous fundamental processes: electro-catalysis, solar energy conversion, energy storage in batteries, and biological events such as photosynthesis. SERS has also been identified as a valuable technique for the detection of explosives/chemical weapons, unmodified DNA, aerosol pollutants and pathogens.

TERS is a technique that provides spectral information with a spatial resolution on the nanometre scale. Since the first reports of TERS emerged in 2000, TERS has become a powerful technique for studying thin crystalline materials, carbon nanotubes, single strands of RNA/DNA, redox reactions, mapping of individual molecules, semi-conductor nanostructures and microcavities.

In the following sections, the fundamental physics that underpins the spontaneous Raman effect, stimulated- and coherent Raman spectroscopy, SERS and TERS are detailed in the context of their applications. Experimental considerations are discussed, and examples of Raman spectroscopy instrumentation setups are presented. The conventions for presenting spectra are detailed and examples of analysis techniques are given for each of the applications of Raman spectroscopy. In the final sections, the recent advances that constitute the current frontiers of Raman spectroscopy are presented from various fields of research worldwide.

Fundamental Principles

When light interacts with matter, the oscillatory electro-magnetic (EM) field of the light perturbs the charge distribution in the matter which can lead to the exchange of energy and momentum leaving the matter in a modified state. Examples include electronic excitations and molecular vibrations or rotational-vibrations (ro-vibrations) in liquids and gases, electronic excitations and optical phonons in solids, and electron-plasma oscillations in plasmas.

Spontaneous Raman

When an incident photon interacts with a crystal lattice or molecule, it can be scattered either elastically or in elastically. Predominantly, light is elastically scattered (i.e. the energy of the scattered photon is equal to that of the incident photon). This type of scattering is often referred to as Rayleigh scattering. The inelastic scattering of light by matter (i.e. the energy of the scattered photon is not equal to that of the incident photon) is known as the Raman effect. This inelastic process leaves the molecule in a modified (ro-)vibrational state. In the case of a crystal lattice, the energy transfer creates a quantum of vibration in the lattice known as a phonon (a quasi-particle). Raman scattering in crystals can also lead to paramagnetic ions, surface plasmons and spin waves. The shift in angular frequency of the scattered light can be described by the following equation:

$$\omega_{\text{scat}} = \omega_{\text{p}} \pm \omega_{\text{osc}},$$

where subscripts osc denotes the lattice or molecule vibration, p denotes the incident photon (often referred to as the pump photon) and scat denotes the scattered light. The binary operator (±) is determined by energy conservation. When the energy of the scattered photon is lower than that of the incident photon (i.e. red shifted), the process is referred to as Stokes Raman scattering. Conversely, when the energy of the scattered photon is higher than that of the incident photon (i.e. blue shifted), the process is referred to as anti-Stokes Raman scattering. The Raman process must also conserve momentum, which is expressed in wave vector form as:

$$\vec{k}_{\text{scat}} = \vec{k}_{\text{p}} \pm \vec{q},$$

where $\vec{k}_{\text{scat}}, \vec{k}_{\text{p}}$ and $\vec{q}$ are the wave vectors of the scattered light, the incident light and the phonon or molecular (ro-)vibration, respectively.

In molecules and crystals, the charge distribution has an equilibrium state to which it tends. An externally applied field can modify or perturb the charge distribution but only in accordance with the molecule or crystal's ability to form dipoles which may be anisotropic. This anisotropic property of molecules and crystals is called the polarisability and dielectric susceptibility, respectively. The classical approach theorises that the existence of the Raman effect is associated with the modulation of the polarisability (for molecular (ro-)vibrations) or dielectric susceptibility (for crystal lattice vibrations) due to the oscillatory nature of their interatomic displacement. For crystal lattice vibrations, consider the polarisation vector of the material, $\vec{P}$. If the suffixes j and k represent the vector components in the x, y and z directions, the jth component of $\vec{P}$ (to first-order) is related to the oscillatory electric field vector $\vec{E}$ associated with the light by:

$$P_j^{(1)} = \varepsilon_0 \, \chi_{jk}^{(1)} E_k,$$

Where ε_0 is the permittivity of free space, χ_{jk} is the dielectric susceptibility of the material (a rank two tensor) and the convention of summation over repeated indices is implied; the superscript (1) signifies that this is the first-order contribution to the polarisation. The polarisability tensor is a function of the nuclear coordinates which, by extension, means that it will also depend on the (ro-) vibrational frequency. Assuming the modulation is small, the dependence can be expressed in a Taylor series with respect to the coordinates of vibration as follows:

$$\chi_{jk}^{(1)}\left(\vec{k}_{\text{p}}, \omega_{\text{p}}\right) \approx \chi_{jk}^{(1)}\left(\vec{k}_{\text{p}}, \omega_{\text{p}}\right)_{\vec{u}=0} + u_l \left(\frac{\partial \chi_{jk}^{(1)}\left(\vec{k}_{\text{p}}, \omega_{\text{p}}\right)}{\partial u_l} \right)_{\vec{u}=0} + u_l u_m \left(\frac{\partial^2 \chi_{jk}^{(1)}\left(\vec{k}_{\text{p}}, \omega_{\text{p}}\right)}{\partial u_l \partial u_m} \right)_{\vec{u}=0} + \ldots,$$

Where $\vec{u}$ is the nuclear displacement vector, the indices j, k, l and m indicate different spatial coordinates with repeated indices in any of the terms implying the summation of the constituents of that index. If we write the electric field associated with the light as follows:

$$\vec{E}(\vec{r},t) = \vec{E}\left(\vec{k}_{\text{p}}, \omega_{\text{p}}\right)\cos\left(\vec{k}_{\text{p}} \bullet \vec{r} - \omega_{\text{p}} t\right),$$

and the nuclear displacement as follows:

$$\vec{u}(\vec{r},t) = \vec{u}\left(\vec{q}, \omega_{\text{osc}}\right)\cos\left(\vec{q} \bullet \vec{r} - \omega_{\text{osc}} t\right),$$

An explicit expression for time dependence of $P_j^{(1)}$ can be found by substitution of these two mathematical equations of the monochromatic light and displacement. The numerous resulting terms pertain to optical processes such as Rayleigh scattering, optical absorption and Raman scattering. The term which pertains to the first-order Raman scattering is derived from the second term on the right-hand side of equation $\chi_{jk}^{(1)}\left(\vec{k}_p,\omega_p\right) \approx \chi_{jk}^{(1)}\left(\vec{k}_p,\omega_p\right)_{\vec{u}=0} + u_l\left(\frac{\partial\chi_{jk}^{(1)}\left(\vec{k}_p,\omega_p\right)}{\partial u_l}\right)_{\vec{u}=0} + u_l u_m\left(\frac{\partial^2\chi_{jk}^{(1)}\left(\vec{k}_p,\omega_p\right)}{\partial u_l \partial u_m}\right)_{\vec{u}=0} + \ldots$, and yields:

$$P_j\left(\vec{r},t,\vec{u}\right) = \frac{1}{2}\varepsilon_0\left(\frac{\partial\chi_{jk}^{(1)}\left(\vec{k}_p,\omega_p\right)}{\partial u_l}\right)_{\vec{u}=0} u_l\left(\vec{q},\omega_{osc}\right)E_k\left(\vec{k}_p,\omega_p\right)\times\left\{\cos\left[\left(\vec{k}_p+\vec{q}\right)\bullet\vec{r}-\left(\omega_p+\omega_{osc}\right)t\right]\bullet+\cos\left[\left(\vec{k}_p-\vec{q}\right)\bullet\vec{r}-\left(\omega_p-\omega_{osc}\right)t\right]\right\}$$

This term contains sum (anti-Stokes) and difference (Stokes) frequencies and demonstrates conservation of momentum as per equations $\omega_{scat} = \omega_p \pm \omega_{osc}$, and $\vec{k}_{scat} = \vec{k}_p \pm \vec{q}$.

The quantum mechanical description of the Raman process states that the (ro-)vibrational energy of the molecules/phonons are discrete quanta. Figure shows an energy level diagram illustrating the Raman processes with Stokes emission at ω_S and anti-Stokes emission at ω_{AS}.

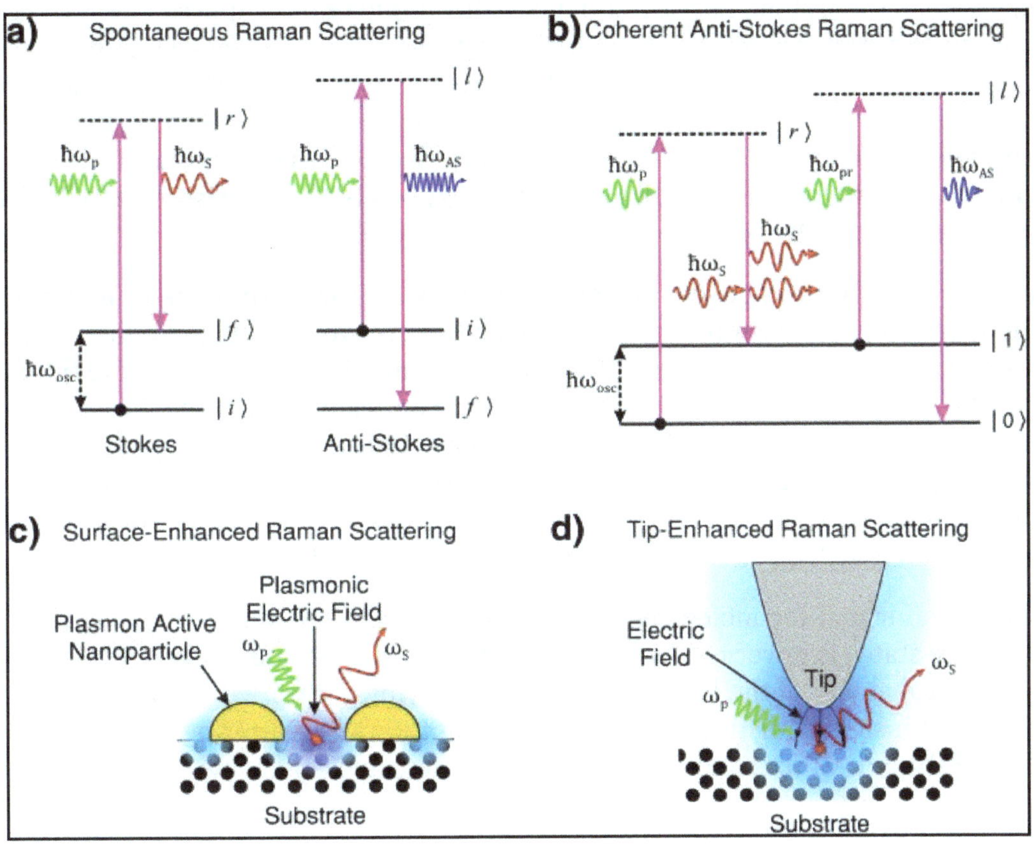

In figure, a) Energy transfer process in Stokes (left) and anti-Stokes (right) Raman scattering, in both scattering processes, the lifetime of the excited state is probabilistic and spontaneous. In Stokes Raman scattering, the initial (ro-)vibrational energy $|i\rangle$ of the scattering material is less than that of the final state $|f\rangle$, the scattered light has less energy than the pump light. In anti-Stokes scattering, the initial (ro-)vibrational energy $|i\rangle$ of the scattering medium is greater than that of the final state $|f\rangle$, the scattered light has more energy than the pump light. b) Coherent anti-Stokes Raman scattering (CARS). CARS is a four-wave mixing process of pump, Stokes, probe and anti-Stokes light in which the emission of anti-Stokes light is coherently induced through an intermediate (ro-)vibrational energy state population inversion. c) Surface-enhanced

Raman scattering (SERS). The incident pump light induces a surface plasmon resonance. The resultant enhancement of the oscillatory electro-magnetic (EM) field strength (shown in blue) on the surface intensifies the light-matter interaction and consequently increases the intensity of the Raman scattered light. d) Tip-enhanced Raman scattering (TERS). The incident pump light induces a tip-surface plasmon resonance associated with the plasmonically active tip. The resultant enhancement of the oscillatory EM field strength (shown in blue) is localised to the vicinity of the tip apex. The lighting rod effect (illustrated by curved black arrows) intensifies the light-matter interaction in the tip region and provides high-resolution (beyond the diffraction limit of light).

In Raman scattering, the intermediate states of the perturbation imposed by the incident pump photon ($|r\rangle$ and $|l\rangle$ in Figure) generally do not correspond to electronic states of the system and are said to be virtual energy states. These virtual intermediate states do not represent a well-defined energy state of the system. As the frequency of the pump photon approaches the energy of the electronic states, the strength of the Raman Effect increases due to resonance effects and is termed pre-resonance Raman. If the intermediate state corresponds to a discrete electronic energy state, the interaction is described as resonance Raman scattering and the signal strength are expected to exceed that of virtual-intermediate-state Raman scattering by orders of magnitude. If the energy of the incident light is in the range of dissociative energy levels, the process is described as continuum resonance Raman scattering.

Raman scattering transitions between certain quantum states are forbidden. In materials with inversion symmetry (i.e. centrosymmetric crystal structure), the initial and final states must have the same parity and are mutually exclusive with absorptive transitions (optically active transitions). In other words, transitions can be either Raman active or optically active. For linear molecules, the symmetric stretching modes of vibration or bending are Raman active and are optically inactive; those with anti-symmetric modes are Raman inactive and optically active (i.e. mutually exclusive). This rule is general and for nonlinear molecules, mutual exclusion is relaxed. In materials without inversion symmetry, (ro-)vibrational mode transition can be both Raman and optically active.

The Stokes Raman signal for molecules is more intense than the anti-Stokes signal as the population of energy states is governed by thermal statistics. For bosonic systems, such as phonons in crystals, the probability of the scattering target occupying a given vibrational quantum energy state obeys Bose-Einstein statistics. Under non-resonant Raman scattering and thermal equilibrium, the ratio of the anti-Stokes and Stokes scattered intensity is given by:

$$\frac{I_{\mathrm{AS}}}{I_{\mathrm{S}}} = \left(\frac{\omega_{\mathrm{p}} + \omega_{\mathrm{osc}}}{\omega_{\mathrm{p}} - \omega_{\mathrm{osc}}}\right)^4 e^{\left(-\frac{\hbar\omega_{\mathrm{osc}}}{kT}\right)}$$

where I_S and I_{AS} are the intensity of the Stokes and anti-Stokes light, respectively, $\hbar$ is Planck's constant divided by 2π, k is the Boltzmann constant and T is the temperature associated with the scattering species. This equation is sometimes used to measure the temperature via Raman spectroscopy. This relation becomes inaccurate for resonance Raman scattering because the Stokes and anti-Stokes processes occur at different pump photon frequencies.

In the case of spontaneous Raman scattering, the Raman Effect is very weak; typically, 1 in 10^8 of the incident radiation undergoes spontaneous Raman scattering. The transition from the virtual excited state to the final state can occur at any point in time and to any possible final state based on probability. Hence, spontaneous Raman scattering is an incoherent process. The output signal power is proportional to the input power, scattered in random directions and is dependent on the orientation of the polarisation. For example, in a system of gaseous molecules, the molecular orientation relative to the incident light is random and hence their polarisation wave vector will also be random.

Furthermore, as the excited state has a finite lifetime, there is an associated uncertainty in the transition energy which leads to natural line broadening of the wavelength as per the Heisenberg uncertainty principle $(\Delta E \Delta t \geq \hbar / 2)$. The scattered light, in general, has polarisation properties that differ from that of the incident radiation. Furthermore, the intensity and polarisation are dependent on the direction from which the light is measured. The scattered spectrum exhibits peaks at all Raman active modes; the relative strength of the spectral peaks are determined by the scattering cross-section of each Raman mode. Photons can undergo successive Rayleigh scattering events before Raman scattering occurs as Raman scattering is far less probable than Rayleigh scattering.

Nonlinear Susceptibility

The polarisation described by equation $P_j^{(1)} = \varepsilon_0 \chi_{jk}^{(1)} E_k$, is in agreement with first-order (i.e. linear) optics and describes the single-photon scattering process (two-wave mixing process). In wave mixing processes with more than two waves, nonlinear optical polarisation must be considered due to the products of the mixed electric field components. Nonlinear optical polarisation can be described by the following:

$$P_j = \varepsilon_0 \left[\chi_{jk}^{(1)} E_k + \chi_{jkl}^{(2)} E_k E_l + \chi_{jklm}^{(3)} E_k E_l E_m + \ldots \right],$$

where $\chi^{(2)}$ is the second-order susceptibility (rank three tensor), $\chi^{(3)}$ is the third-order susceptibility (rank four tensor) and the sum over repeated subscript indices is again implied. Each of the terms in Eq. above can be written in shorthand by $\vec{P}^{(1)}$, $\vec{P}^{(2)}$, $\vec{P}^{(3)}$, etc. The physical processes that occur because of the second-order polarisation, $\vec{P}^{(2)}$, tend to be distinct from those arising from the third-order polarisation, $\vec{P}^{(3)}$. This polarisation can have electric dipole, quadrupolar, octupolar, (etc.) contributions. Under the electric dipole approximation, the second-order polarisation can only occur in crystals that are noncentrosymmetric (lack inversion symmetry). Hence, $\chi^{(2)}$ vanishes for media such as fluids (e.g. liquid/gas) and amorphous solids (e.g. glass). Third-order nonlinear optical interactions (i.e. those described by a $\chi^{(3)}$ susceptibility) can occur for both centrosymmetric and noncentrosymmetric systems. Electric quadrupolar, octupolar, (etc.) $\chi^{(2)}$ contributions do not disappear under inversion symmetry.

Stimulated Raman Scattering

While spontaneous Raman scattering is an incoherent process, SRS is a coherent four-wave nonlinear optical mixing process. The modes of oscillation are in phase forming a coherent modulation of polarisation in the sample with susceptibility $\chi^{(3)}(\omega_S; \omega_p + \omega_S - \omega_p)$. The scattered light is also coherent. The SRS process is dependent on the spontaneous Raman cross-section, the spectral linewidth, the path length of the light-field-matter interaction, the input intensity and optical feedback (light generation) of Stokes frequency light.

When photons of frequency ω_p and ω_S simultaneously interact with a molecule or crystal lattice in the ground state, the system vibrates with an induced frequency: $\omega_{osc} = \omega_p - \omega_S$. Unlike spontaneous Raman scattering, the de-excitation (relaxation) time to and energy of the final state are determined by the stimulation effect. The interaction results in the transfer of energy from the pump photon to the molecule/lattice, and the molecule/crystal scatters a new photon with frequency and phase matching that of the incident light of frequency ω_S. Figure shows the process schematically.

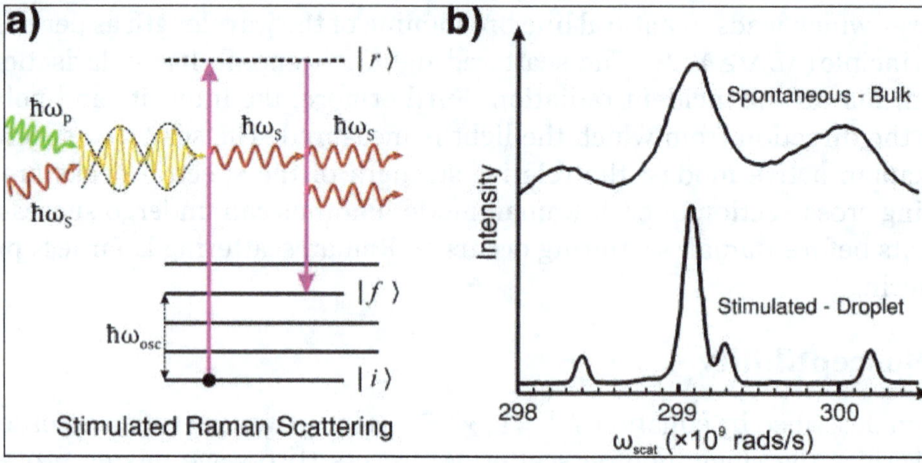

In figure, a) Energy level diagram of stimulated Raman scattering (SRS). SRS is the induced emission of Stokes light by the coherent interaction of the pump and Stokes light with the material.

Unlike spontaneous Raman scattering where the lifetime of the state $|r\rangle$ and the energy of the final state $|f\rangle$ are probabilistic, in SRS, the (ro-)vibration of the molecule or lattice is coherently driven by the difference frequency of the pump and Stokes light. b) Comparison of spontaneous Raman scattering and SRS of bulk and droplet ethanol. The spontaneous measurements were performed in a cuvette (bulk ethanol). The SRS measurements were performed in a droplet of ethanol which acted as an optical resonator for the Stokes light.

It is common to employ an external radiation source tuned to the Stokes frequency in tandem with the pump laser beam to provoke this effect. This technique can lead to exponential gain in the Stokes signal, by transferring energy from the pump radiation, and rapid population of the final (ro-)vibrational state $|f\rangle$. However, if the intensity of the incident light of frequency ω_p is sufficient, the generation of Stokes frequency photons within the material can self-promote SRS without the need for an external ω_S source. The intensity threshold of incident light in organic liquids, such as ethanol, for this kind of self-generated SRS typically requires an incident peak intensity of pump light $> 10^9$ W/cm^2 for an optical path length of a few centimetres. However, the SRS threshold can be significantly reduced by extending the length of the pump and Stokes field interaction with an optical resonator, such as internal reflection in a droplet of liquid. The example shown in Figure is the SRS spectrum taken with droplets of ethanol directly compared to the spontaneous Raman spectrum of bulk ethanol. The droplets act to confine the light by internal reflection which feeds back the Stokes light as a self-SRS inducing optical resonator.

Coherent Anti-Stokes Raman Scattering

CARS is a third-order nonlinear four-wave optical mixing process. Figure shows the energy level diagram for the process. A pump beam and probe beam of frequency ω_p and ω_{pr} are mixed with a third beam of frequency ω_S (Stokes frequency) and incident on the sample. The frequency difference $(\omega_p - \omega_S)$ needs to match the frequency associated with the Raman active (ro-) vibrational mode $\omega_{osc} = \omega_p - \omega_S$. The frequency of the Stokes beam is usually adjusted/tuned to satisfy this criterion. Next, a probe photon of frequency ω_{pr} provides a perturbation for the anti-Stokes scattering process to occur at frequency $\omega_{AS} = \omega_p - \omega_S + \omega_{pr}$. A macroscopic third-order polarisation, $P^{(3)}$, is induced due to the coherent superposition of the microscopic dipole oscillations. Hence, CARS is governed by the third-order susceptibility of the form: $\chi^{(3)}(\omega_{AS}; \omega_p - \omega_S + \omega_{pr})$.

There are numerous treatments and approaches to formulating expressions for $\chi^{(3)}$. If one assumes that the excitation field is much weaker than the intramolecular forces, then a perturbative approach can be adopted. If this is not the case, non-perturbative treatments can be considered. By considering the density matrix equation of the system and expressing the external field interaction as a perturbation in the Hamiltonian, the semi-classical nonlinear optics theory generates an expression for $\chi^{(3)}$ with 48 terms, each of which contribute to the third-order susceptibility. A generalised expression for dominant terms in resonant CARS is given by the following:

$$\chi^{(3)} = \frac{A_R}{\omega_{osc} - (\omega_p - \omega_S) - i\tilde{\Lambda}_R} + \chi_{NR}^{(3)},$$

Where Γ_R is the half width at half maximum for the Raman line; A_R is a constant representing the Raman scattering cross-section. The first term is the contribution due to CARS vibrational resonance as in Figure, $(\omega_{osc} = \omega_p - \omega_S)$. The second term is the nonresonant background signal and is independent of the Raman shift $(\omega_{osc} \neq \omega_p - \omega_S)$. The nonresonant background occurs because not all quantum pathways of the scattering process involve a resonance with a (ro-)vibrational state. This nonresonant contribution interferes with the resonant part of the signal. The nonresonant background causes distinctive distortions of CARS spectra in comparison with spontaneous Raman spectra and has prevented CARS from becoming a widespread technique.

The incident light beams of differing frequency move in and out of phase with each other in both time and space. Hence, the CARS signal reaches its first maximum when the field-sample interaction length scale is less than the coherence length scale to yield constructive interference. For *plane-wave* pump and Stokes beams, the intensity of the anti-Stokes signal is as follows:

$$I_{AS} \propto \left|\chi^{(3)}\right|^2 I_p I_{pr} I_S \left(\frac{\sin\left(\Delta\vec{k} \bullet \frac{\vec{z}}{2}\right)}{\frac{\left|\Delta\vec{k}\right|}{2}} \right)^2,$$

Where $\vec{z}$ is the sample thickness (vector normal to the lattice cell surface), $\vec{k}$ is the wavevector of light, $\Delta\vec{k} = \vec{k}_p - \vec{k}_S + \vec{k}_{pr} - \vec{k}_{AS}$ is the wavevector mismatch (the velocity difference between the four waves) and I_i is the intensity of the wave denoted by the subscript. Phase matching is achieved when $\Delta\vec{k} = 0$ and the intensity of the anti-Stokes signal is maximised because the energy and momentum transfer processes correspond to allowed transitions. As the magnitude of $\chi^{(3)}$ is linearly proportional to the number oscillators involved in the process, the intensity of the anti-Stokes signal is quadratically proportional to the number/concentration of oscillators.

Researchers typically employ the pump beam to provide the second virtual excitation (i.e. the probe light shown in Figure; i.e $\omega_{pr} = \omega_p$ and $\omega_{AS} = 2\omega_p - \omega_S$). The intensity of the CARS signal is therefore quadratically proportional to the intensity of the pump beam. The CARS signal is mono directional due to the phase-matching condition. However, high numerical aperture (NA) lenses or microscope objectives (confocal light) are commonly employed to satisfy the phase-matching condition without the need for complex mechanical systems to achieve phase-matched beam alignment.

Raman resonances typically have coherence times of ~ 1 ps. Hence, the pump and Stokes beams are typically pulsed in picosecond timescales to obtain coherent excitation and to inhibit multiphoton effects. The CARS process takes place in the immediate vicinity of the beam's focal spot. The signal produced is typically 10^6 times that of spontaneous Raman scattering. CARS microscopy offers non-invasive characterisation and imaging of (ro-)vibrational spectra with high sensitivity and spectral resolution as well as three dimensional sectioning capabilities.

Surface Plasmons and Polaritons

Surface plasmons can occur at the interface between a dielectric and conducting material, such as a metal or degenerate semi-conductor. They are the light-induced coherent oscillations of surface conduction electrons about their equilibrium position. The nanoscale volume of opposing charge that remains acts as a restoring force on the electrons. The result can be described with a damped simple harmonic oscillator model, in which the oscillations of the free-charge carriers have an associated resonance. Surface plasmons can be excited by EM radiation and plasmonics is the study of these light-matter interactions.

Plasmonic nanoparticles that are much smaller than the wavelength of the incident light can support non-propagating surface plasmons that oscillate with a frequency known as the local surface plasmon resonance (LSPR). The wavelength of surface plasmons is much shorter than that of the associated propagating light for a given frequency. The LSPR wavelength is dependent on the nanoparticle's size, shape, material, external dielectric properties and inter-nanoparticle separation.

Surface plasmons that propagate are referred to as surface plasmon polaritons (SPPs). They are essentially light waves that are trapped at the interface due to their interaction with the free electrons of the conducting material. For a planar dielectric-conducting interface, polaritons propagate in 2-dimensional space along the surface interface for length scales of tens to hundreds of micrometres. They decay evanescently in the direction perpendicular to the surface interface with 1/e decay lengths of up to 200 nm. The field intensity in this evanescent decay region is amplified by orders of 10 to 100 relative to the incident radiation. Hence, light-matter interactions with adsorbed molecules on the surface are also enhanced.

In the case of LSPR, the surface roughness or surface nanoparticles cause local concentrations of charge carriers which further amplify the evanescent EM field due to the lightning rod effect. Even larger field-enhancements (up to 10^6) can be observed in gap plasmons (in the gap between two neighbouring plasmonic nanoparticles). This enhanced near-field effect gives rise to the technique known as SERS.

Surface-Enhanced Raman Scattering

Raman is generally a very weak process; it is estimated that approximately one in every 10^8 photons undergo Raman scattering spontaneously. This inherent weakness poses a limitation on the intensity of the obtainable Raman signal. Various methods can be used to increase the Raman throughput of an experiment, such as increasing the incident laser power and using microscope objectives to tightly focus the laser beam into small areas. However, this can have negative consequences such as sample photo bleaching. Placing the analyte on a rough metal surface can provide orders of magnitude enhancement of the measured Raman signal, i.e. SERS.

Two mechanisms have been proposed to explain the increase in Raman signal provided by SERS. The first is via EM enhancements where local surface plasmons concentrate the local electric field near the surface of the metal in 'hot spots' located on the sharp edges of nanostructures or in regions of concentrated free-charge carriers due to the lighting rod effect. Figure, illustrates the SERS process. This process can increase Raman generation by a factor of 10^8 to 10^{11}. The second method is chemical enhancement via charge transfer between the metal surface and the analyte, which enhances Raman scattering by a factor of approximately 10^2 to 10^3. However, the charge transfer mechanism only applies to specific molecules, whereas the EM mechanism is applicable for all analytes.

The ubiquity of EM enhancements has led to the development of numerous SERS substrates, which can be divided into two groups: metallic nanostructures fabricated on a solid substrate and colloidal suspensions of plasmonic nanoparticles. The most common materials used to fabricate SERS substrates are gold and silver because of their good plasmonic response. Gold also benefits from chemical stability as it is a noble metal. Other metals are also being investigated, such as aluminium for UV Raman spectroscopy.

Tip-Enhanced Raman Scattering

The diffraction limit of light restricts the focus spot size in standard optical techniques (such as Raman spectroscopy) to be at least half of the wavelength of the light according to Abbe's criterion. Light from the sample is composed of both propagating and non-propagating radiation. The non-propagating evanescent waves remain in the vicinity of their sources and do not participate in image formation in the far field. Instead, they extend laterally on the sample among the plasmon-active sites. Hence the spatial resolution is restricted by the size of the focal spot of the light. Even with a focal spot size of a half-wavelength ($\sim$250 nm for visible light), any objects that are much smaller than the half-wavelength would appear as a defuse shape.

TERS is a relatively new optical nanoimaging technique that combined Raman spectroscopy with scattering (or apertureless) scanning near-field optical microscopy. TERS offers spatial resolution far beyond the diffraction limit of the probing light. In the context of the a priori description, this is achieved by forcing the near-field evanescent light into the far-field image formation. At the present date, the spatial resolution of TERS is typically reported to be 10–30 nm and is largely assumed to scale with the size of the tip's apex. Incremental improvements to this resolution have been reported. Enhancement factors for TERS are significantly weaker than SERS due to the relative size of the probed signal volume. The enhancement factor (relative to spontaneous Raman scattering) is typically reported to be 10^3 to 10^6. As with SERS, two field enhancement mechanisms are thought to contribute to the Raman signal: EM and chemical enhancement.

TERS is implemented by positioning a plasmon-active (plasmonic) nanotip approximately 50 nm above the sample's region of interest. The Raman probe light is focused onto the tip-surface cavity to induce LSPR within the tip' apex and (in some circumstances) the sample surface. The surface plasmons may then enhance evanescent or near-field light with the incident probe light and/or the Raman scattered light. Hence, the LSPRs both confine and enhance the light field in the vicinity of the tip's apex. The enhanced local EM field is most concentrated at the tip apex due to the lightning rod effect. This evanescent light at the tip apex can then excite or stimulate Raman, two-photon or second harmonic scattering from a nanoscale volume of the sample under the tip. A Raman image of the sample surface can be obtained by raster scanning the sample under the nanometric tip.

Experimental Considerations

Instrumentation

The non-resonant Raman Effect is a very weak process. Hence, monochromatic, narrow-beam and high-intensity lasers are preferable to produce quality Raman spectra. The exploitation of microelectronics, such as stepper motor drives, photon counters, digital data acquisition and computational processing systems can further enhance the quality of spectra. As spontaneous Raman spectroscopy is naturally an incoherent process, continuous-wave laser sources are commonly used because pulsed lasers require higher peak powers for sufficient signal-to-noise ratio, which can photo bleach/damage samples.

The choice of wavelength of the laser source depends on the required application. Lower visible wavelengths and UV cause strong photoluminescence in organic materials, which can mask the Raman peaks. Therefore, a longer visible or near-IR wavelength (500—830 nm) laser source is often suited for studying organic materials, because of the reduced photoluminescence. However, the Raman signal intensity is inversely proportional to the wavelength of the pump light. Hence, longer wavelengths of light require longer acquisition times.

Raman spectroscopy is most often performed using laser sources at $\lambda = 785$ nm. This wavelength source is often selected as it balances the competing factors between Raman signal intensity, fluorescence, detector sensitivity and cost, and cost-effective/compact high-quality laser sources. However, visible lasers in the blue and green (e.g. $\lambda = 532$ nm) are becoming more common in Raman spectroscopy.

Raman scattering is measured in terms of the wavelength shift from the source wavelength. Ideally the illumination source for Raman measurements should be purely monochromatic, in other words, a single wavelength. However, all laser sources possess a spectrum of wavelengths known as a linewidth. The linewidth of a laser is usually measured in Hertz and is typically >1 MHz for solid-state lasers used in Raman applications. A narrow linewidth is preferable for Raman spectroscopy because the measured shift in the Raman scattering process is limited by the laser's linewidth.

Laser sources for Raman spectroscopy need to be stable in wavelength and power over extended periods of time and from use to use. Raman spectra are usually collected over long integration times and for many acquisitions. If the wavelength of the source drifts during a measurement, then the Raman peaks will drift as well, because Raman is measured as a shift relative to the pump light. Wavelength drift is also problematic from measurement to measurement as it causes peaks to shift, in turn making comparisons between measurements difficult. The output power stability of the source is important for similar reasons. If the laser power drifts from measurement to measurement, then quantitative comparisons cannot be made easily.

Spectral purity is another key criterion for Raman laser sources. The spectral purity of laser sources often requires side-mode suppression better than 60 dB. In many cases, side-mode suppression is sufficient if > 60 dB spectral purity is reached at ~ 1–2 nm from the laser wavelength peak. However longer wavelength (near-IR) Raman spectroscopy requires side-mode suppression ratios within a few hundreds of pm from the main peak.

Most modern Raman systems use solid-state laser sources rather than gas lasers because of their spectral quality and stability. There are three main categories of continuous-wave solid-state laser sources used in Raman spectroscopy: Diode-pumped single-longitudinal mode (SLM) lasers; single-mode diode lasers (distributed feedback (DFB) or distributed Bragg reflection (DBR)); and volume Bragg-grating (VBG) frequency-stabilised diode lasers. These laser sources have varying optical characteristics.

Diode-pumped SLM lasers are readily available in compact form from the UV to the near-IR. Power levels of several Watts are achievable at 1064 nm in the near-IR. In the visible range, numerous lines in the blue-green-red region (457 to 660 nm) are available with output powers of ~ 100 mW. In the UV spectral range, power outputs of 10–50 mW at 355 nm are available. Hermite-Gaussian laser beam modes are described by their transverse electro-magnetic mode (TEM): $TEM_{m,n}$, where m and n represent the Hermite-Gaussian mode index. Diode-pumped SLM lasers provide excellent TEM_{00} mode beams, precise wavelengths with low drift, and a single-frequency linewidth > 1 MHz. The spectral purity of diode-pumped SLM lasers is typically > 60 dB in terms of their side-mode suppression ratio. Weak emissions that neighbour the laser's main peak several nanometres in spectral shift can occur in diode-pumped SLM lasers. However, these neighbouring lines can be mitigated with dielectric band-pass filters. The wavelength of diode-pumped SLM lasers is typically stable to within 4 pm over a temperature change of 30 °C.

Single-mode diode lasers are compact and cost-effective pump illumination sources with single-frequency linewidth (> 1 MHz), single-TEM beam quality and output powers of up to ~ 100 mW. Wavelengths of λ = 785, 830, 980 and 1064 nm are most common in Raman spectroscopy. The side-mode suppression ratio is typically limited by sideband emission to ~ 50 dB at ~ 100 pm from the main peak.

VBG frequency-stabilised diode lasers use a narrow-linewidth VBG element with a diode-laser emitter to achieve narrow-line emission. These lasers are often used for applications requiring narrow-line emission at wavelengths that are not available for DFB or DBR laser sources. Frequency-locking multi-TEM diode lasers can be used to increase the output power of the narrow-linewidth emission. The stability of the output wavelength and linewidth requires careful thermomechanical control and high-precision alignment inside VBG frequency-stabilised diode lasers. Linewidths can range from single-frequency emission to ~10s of pm, depending on the wavelength and the output power. The side-mode suppression ratio is limited to ~50 dB, ~250 pm from the main peak emission. However, this can be improved using filters.

In confocal Raman imaging applications, it is necessary to use diffraction-limited TEM_{00} beams for optimum spatial resolution. However, this is relaxed for probe-based quantitative Raman analysis. In addition, confocal Raman setups require laser beam isolation as samples may generate optical feedback that is well aligned to the incident pump light. This counter-propagating feedback can induce power and noise instability and can even damage the laser source. Optical isolators are often integrated into the laser system itself because careful alignment must be achieved in the output after the isolator.

The spectrometer is a core component of any set-up used for measuring Raman spectra. The spectrometer should match the wavelength(s) of the laser source(s) used. The spectral range and resolution required will depend on the application. For example, the spectral range is determined by the position of the Raman peaks of interest (i.e. at large $\Delta\tilde{\nu} \sim 300\,cm^{-1}$ or low $\Delta\tilde{\nu} \sim 1\,cm^{-1}$). If the application requires closely spaced Raman peaks to be resolved, then spectral resolution is key. The spectral resolution of a spectrometer is largely determined by the slit width at the spectrometer entrance, the focal length of the spectrometer, the dispersion, the size of the grating (or prism) and the size and sensitivity/quality of the detector. There is a trade-off between the overall spectral range and resolution when considering the design of the experiment for a given application. In the case of weak Raman signals, optimising the signal-to-noise ratio is a priority.

Spectral filtering plays a vital role in the acquisition of Raman spectra. Firstly, the incident laser light must be spectrally pure, which is accomplished with a narrow-linewidth laser source. However, if the laser light is delivered to the sample by an optical fibre, then it is inevitable that Raman generation will occur in the fibre. Therefore, it is important to use a narrow band-pass filter to reject any Raman signal generated in delivering the laser to the sample. Narrow band-pass filters can provide transmission >90 % at the laser wavelength while suppressing light to an optical density of OD > 5 at wavelengths differing by just 1% from the laser wavelength.

Importantly, light collected for detection requires filtering to block the laser wavelength. If the laser light is not filtered out, it can go on to generate Raman in the detection arm of the set-up and drown out the desired Raman signal when it reaches the spectrometer. The type of filter required depends on whether Stokes, anti-Stokes or both are to be measured. To only detect anti-Stokes Raman, a short-pass filter should be used as anti-Stokes Raman light has a higher energy and hence shorter wavelength than the laser source. To only detect Stokes Raman, a long-pass filter should be used as the Stokes Raman light has a lower energy and hence longer wavelength than the laser source. Long pass edge filters with edge-transition widths of <3 nm and edge steepness <40 cm^{-1} are available. To detect both Stokes and anti-Stokes Raman light, a notch filter centered on the laser wavelength should be used as it allows both shorter and longer wavelengths to be detected. Notch filters with OD > 6 at the laser line wavelength are available. Multi-notch filters are also available and can block multiple laser lines simultaneously. Holographic notch filters significantly outperform dielectric notch filters, providing excellent attenuation of the Rayleigh line while passing light as near as 50 cm^{-1} from the Rayleigh line. Acousto-optic modulators can also be used in conjunction with an excitation laser to select emissions with a desired wavelength (as a filter) or as a time-gated illumination system in tapping mode atomic force microscopy (AFM)-based TERS.

The quantum efficiency of standard room-temperature silicon-based CCD devices for Raman signal detection degenerates above λ = 800 nm. For longer wavelengths, indium gallium arsenide array devices can be used, but these are less sensitive with higher noise levels and cost. The visible to near-infrared wavelength range (λ = 500–830 m) is particularly suitable for inorganic materials (e.g. graphene, carbon nanotubes (CNTs) and fullerenes) and SERS. UV lasers are attractive for organic materials (e.g. pathogens, proteins, DNA, and RNA). For materials with strong fluorescence that require near-IR illumination, it is common to use a 1064-nm wavelength.

Spontaneous and Coherent Raman Scattering Setups

Spontaneous Raman spectroscopy is most commonly used for modes with forbidden single-photon absorption or emission experiments. SRS is sometimes used for wavelength shifting of coherent light, light amplification, pulse compression, phase conjugation and beam combining. Unlike spontaneous Raman scattering, SRS is highly directional and offers enhanced signal strength and the ability to time-resolve the evolution and dephasing of coherent (ro-)vibrational motion.

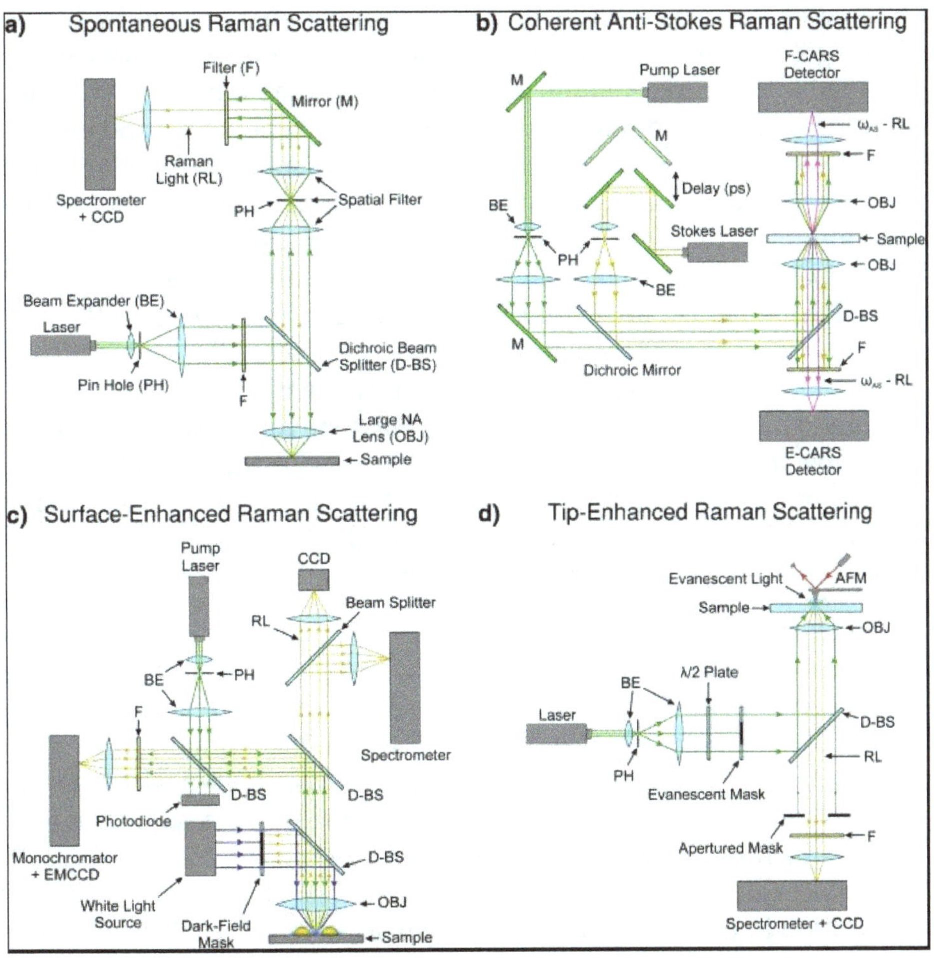

Figure shows a typical Raman setup based on a confocal geometry used by Wiedemeier et al. Confocal setups of this type are commonly used and employ an infinity-corrected objective lens (large numerical aperture (NA) lens) to focus the pump light. Wiedemeier et al. used a diode-pumped solid-state laser as a monochromatic light source centered at 532 nm. Confocal mode is achieved by the use of a pinhole module in front of the spectrometer to spatially filter the light. The pinhole only passes light that originates from the focal plane to the detector. For detection of the Raman signal, a holographic-imaging spectrometer with an attached CCD camera is used. A holographic transmission grating with high light throughput served as a dispersive element, which enables large spectral ranges in a comparatively short time period to be acquired. Raster scanning of the sample in a confocal setup needs to be precise. Hence, a piezo actuated nano-positioner is used for positioning of the specimen.

In figure, a) Typical confocal Raman spectroscopy setup. The pump laser is spatially filtered through a pinhole. The back-scattered Raman light is spatially filtered and spectrally filtered through a notch filter. The Raman light is analysed by a spectrometer and a charge-coupled device (CCD). Hyperspectral images are obtained by raster scanning the sample. b) Typical CARS setup. Two laser sources provide the pump and Stokes light and are synchronised through a picosecond path difference mirror setup. In this setup, the incident light is focused through an optically transmissive sample substrate. Both the forward scattered light (F-CARS) and epi-scattered light (E-CARS) are spectrally filtered by band-pass filters and are subsequently detected by two avalanche photodiodes. CARS images are obtained by raster scanning the sample. c) Typical SERS setup. The pump laser is coupled into a dark-field microscope in which the Raman light is edge-filtered and detected through a monochromator and EMCCD. The white-light source and dark-field mask provides the means for dark-field spectroscopy. The dark-field spectra of each plasmonically active nanoparticle are recorded through a secondary spectrometer (top right in c). An imaging CCD camera is used to automatically find and centre each nanoparticle. d) Typical TERS setup. The pump laser light is spatially filtered and passed through a half-wave plate. The evanescent mask ensures that only high numerical aperture (NA) pump light is incident on the sample such that total internal reflection occurs at the substrate-sample interface. This ensures that the tip apex is only illuminated by the evanescent light to achieve nanoconcentrated light in the vicinity of the tip. The reflected Raman light is filtered by an apertured mask (to remove any residual large NA pump light) and a notch filter. The Raman light is analysed by a spectrometer and a CCD. Hyperspectral images are obtained by raster scanning the sample. F, filter; M, mirror; RL, Raman light; CCD, charge-coupled device; PH, pinhole; BE, beam expander; D-BS, dichroic beam splitter; OBJ, Large numerical aperture (NA) lens; EMCCD, electron-multiplying charge-coupled device.

Spontaneous anti-Stokes scattering is weaker than Stokes Raman scattering due to the relatively low probability of thermal excitation. Hence, anti-Stokes Raman spectroscopy is typically used with stimulated or coherent spectroscopy. CARS spectroscopy offers a 10^5 increase in conversion efficiency, spectral and spatial discrimination against fluorescence and, most importantly, does not require a monochromator. Due to the required coherence of the process, high-peak power pulsed tuneable laser sources are employed. These peaks are readily available using picosecond or femtosecond light lasers, the choice of which is determined by the spectral resolution required and the timescale of interest. Avoiding direct electronic excitations in the sample is an important consideration as photochemical damage (due to photobleaching) can occur in samples. Djaker et al., for example, use near-infrared laser sources to mitigate photobleaching in their samples of polystyrene beads.

Figure shows a typical CARS setup that measures both forward scattered light (F-CARS) and back- or epi-scattered light (E-CARS). The system has two synchronised picosecond pulse trains. The pump and Stokes beams are generated by two picosecond Ti:Sapphire lasers operating at 80 MHz and are tuneable from 700 to 1000 nm to cover the entire spectrum of molecular (ro-)vibrations in biological systems (up to $\Delta\tilde{\nu} \sim 3000\ cm^{-1}$). The ps pulse duration is adjustable by a Gires-Tournois interferometer. The Ti:Sapphire lasers are pumped by a frequency-doubled CW Nd:Vanadate laser that provides monochromatic light at 532 nm. The two pulse trains were polarised with a pulse duration of 3 ps, corresponding to a spectral width of $1.76\ cm^{-1}$. The pump and Stokes beams are synchronously pulse picked through two Bragg cells to reduce the repetition rate of the pulse trains to several hundred kilohertz, thus avoiding photodamage of the sample while still maintaining high-peak power for CARS generation. The pump and Stokes beams are temporally synchronised by a SynchroLock system, which electronically adjusts the time delay between the two pulse trains. A small part of the output of the lasers are launched in optical fibres coupled to photodiodes and connected to a SynchroLock controller, which measures the lasers frequency or phase difference between the master and the slave; the timing jitter was reported to be ~ 250 fs. The spectral resolution was estimated to be $2.5\ cm^{-1}$, which is high enough to resolve Raman spectral features of biological samples. The use of a broadband Stokes wave enables the acquisition of a full CARS spectrum in only one measurement, with this configuration being known as multiplex or broadband CARS.

The two pulse trains are spatially filtered, collinearly combined and expanded through beam expanders. They are then sent into an inverted microscope and focused onto the sample by a water-immersion objective lens with a large NA. The E-CARS signal is collected by the same objective lens while the F-CARS signal is collected by a condenser lens with a lower NA. The E-CARS and F-CARS signals are filtered through a set of band-pass filters and detected by two avalanche photodiodes with a 200 μm × 200 μm active area. The CARS images are collected by raster scanning the sample, using an XYZ piezo flexure stage.

Several methods have been developed to suppress the nonresonant background associated with CARS. E-CARS are relatively insensitive to the nonresonant background of sample solvents. Polarisation-sensitive CARS can differentiate the resonant and nonresonant signals by their polarisation. However, these two techniques reduce the anti-Stokes signal strength. Time-resolved CARS, temporal or spectral interferometry CARS and frequency-modulated CARS can also suppress the nonresonant background. However, the setup in terms of both optics and electronics is challenging.

SERS Specific Considerations

A variety of nanostructures, such as bowtie antennas, nano-rings, nanovoids, nanoparticle aggregates, nanoflower, nanorod arrays and nanowells can be used for SERS. Each nanostructure can have a number of plasmonic resonances, and matching the excitation laser to these wavelengths can greatly enhance the SERS intensity. Matching the plasmonic resonance to the pump laser can be done either by tuning the laser wavelength or by tuning the LSPR of the nanostructures.

The difficulty faced in producing SERS substrates is consistency in fabrication and repeatability in measurements due to the inhomogeneity and randomness of SERS active hot spots. For SERS substrates produced by top-down methods, such as electron beam lithography, the main challenge is scaling the fabrication. Conventional top-down methods limit the active area of the SERS substrate and are not conducive to large-area manufacturing. Bottom-up fabrication methods have their own set of problems. Even though bottom-up approaches allow wafer scale fabrication, consistency across the wafer is usually lacking. This inconsistency hinders the repeatability of measurements, which is problematic for quantitative analysis. Colloidal SERS schemes suffer from complications introduced by stabilising agents at the surface of the nanoparticles, which help to keep nanoparticles in suspension. These stabilising agents can either impede or augment the measured Raman signal. The chemical synthesis for nanoparticle colloids also requires precise optimisation. The poor reproducibility of nanoparticle colloidal synthesis hampers batch-to-batch consistency.

Often, only very few sites exhibit the highest SERS enhancement and the variability in size and shape can alter the plasmonic properties from the desired LSPR. Figure, shows a setup which combines SERS with dark-field spectroscopy. The dark-field spectrometer analyses the light scattered from the nanostructures (illuminated by the white-light source) to select nanostructures with the desired plasmonic properties.

TERS Specific Considerations

Scanning probe microscopy (SPM) techniques, such as atomic force microscopy (AFM), scanning tunnelling microscopy (STM) or shear force microscopy (SFM), are usually the tools of choice for TERS. TERS has the ability to simultaneously measure topography by the conventional SPM mode of the system and obtain corresponding spectral information from a sample with nanometric spatial resolution and high sensitivity. Certain SPM techniques ordain probe modifications for the plasmonically induced nanoscale evanescent light to activate/enhance the Raman signal. The tips can either be made of a metal or coated with a thin layer of metal to modify them for TERS. When the apex of a metallic or a metal-coated nanotip is illuminated with focused light at the LSPR wavelength, local surface plasmons around the tip apex are excited, and evanescent light is produced at the tip apex. This evanescent light can generate Raman scattering from a sample placed right under the tip apex.

The process of Raman scattering takes place in the near-field and the spectral signal is scattered and converted back to the far-field by the tip apex, which is then collected by the usual optics and spectrometer in the far-field. Figure shows such a TERS setup with a modified AFM. The setup consists of largely similar equipment shown in Figure. An inverted microscope illuminates the sample from underneath and the tip is placed at the top surface of the sample. The Raman back-scattered signal is then directed to the spectrometer. An evanescent mask blocks the central part of the laser beam inhibiting the low NA component of the incident light, so that only the high-NA component of the incident light reaches the sample so that total internal reflection occurs. This limits the transmitted light that falls onto the tip and, hence, only the evanescent light participates in the Raman scattering signal. Suppressing the participation of transmitted far-field light reduces the unfavourable background signal.

Polarisation-dependent TERS can be performed with light polarisation parallel to the tip apex in addition to the in-plane linear and radial polarisations. Polarisation dependent TERS is enabled by the large incidence angle from the high-NA objective lens and the use of devices that modify the polarisation state of the light such as a $\lambda/2$ waveplate. The Raman scattered light is then collected in the low NA region through an apertured mask, which inhibits any residual laser light. As the tip apex approaches the sample within the focal spot, evanescent light is created at the tip's apex. Since the intensity distribution within laser focus is not uniform, it is very important to lock the relative position of laser focus to the tip.

The strength and resolution of TERS depends on the ability of the tip to enhance and confine the light field at the tip's apex, respectively. In STM systems, the tips are made of solid metal and the substrates need to be conductive in order to control the tunnelling current. The STM tip resembles a long and smooth nanocone, with an apex diameter of ~20 nm. The length of the tip (~tens of micrometres) makes them plasmonically unfavourable for visible light enhancement. However, the tunnelling gap between the tip and the sample can be tuned to the desired LSPR wavelength, creating a strong hotspot within the gap. Some of the more advanced STM systems allow high-vacuum and low-temperature measurements. As the substrate in STM needs to be conductive (often opaque in the visible wavelength range), the setup shown in Figure. would not be suitable. Hence, a side illumination and side collection configuration is more common with STM-based TERS. To prevent the objective from mechanically interfering with the STM tip, a lens with a long working distance is required. It is therefore not trivial to tightly focus the incident light on the tip apex. A parabolic mirror can be used to mitigate mechanical interference and tightly focus the incident light to the tip apex as well as to collect the Raman signal.

The spatial resolution in TERS is comparable to the size of the metallic nanostructure at the tip apex. The gain in spatial resolution comes at a cost to overall signal enhancement (relative to SERS) due to the reduction of the Raman active volume.

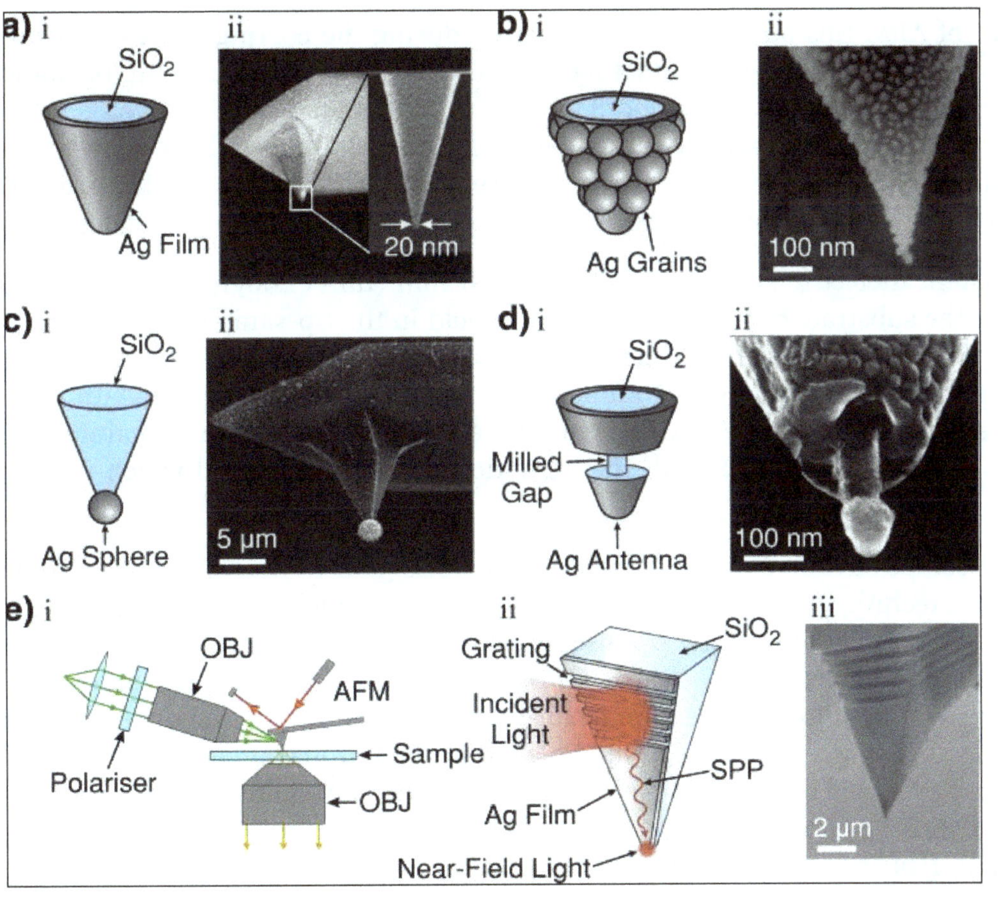

In AFM systems, the tips are usually semiconductor cantilevers, with an apex diameter of ~5 nm. Figure shows five examples of AFM-based TERS tips that have been demonstrated in the literature. The semiconductor tips are usually coated with metal either by thermal evaporation under high-vacuum or electroless metal plating (mirror reaction) techniques. Figure shows an example of a smooth AFM TERS tip. As the substrate does not need to be conductive, AFM-based TERS can be performed in either bottom-up transmissive illumination or in side/top reflective illumination configurations; the transmissive configuration in Figure. is more common.

In figure, a) i, Smooth metallic (silver; Ag) film-coated dielectric (silicon-dioxide; SiO_2) atomic force microscope (AFM) tip. a) ii, SEM image of a Ag-coated AFM tip. After Ag coating by thermal evaporation, a thin granular Ag layer is deposited onto the tip. b) i, Rough Ag-nanoparticle-coated SiO_2 AFM tip. b) ii. SEM image of rough Ag-grain-coated SiO_2 AFM tip formed during the thermal evaporation process. c) i, Single Ag nanoparticle attached to the apex of a SiO_2 AFM tip. c) ii, SEM image of an AFM tip after photoreduction to selectively fabricate an Ag nanoparticle at the tip apex. d) i, Ag-coated SiO_2 AFM tip with a focused ion beam (FIB) milled gap. d) ii, SEM image of antenna fabricated by FIB milling of annular ring and subsequent Ag thermal evaporation from under the tip. The mushroom shape shadows the annular ring from Ag coating. e) i, Illustration of side illumination TERS for surface plasmon polariton (SPP) nanofocusing. OBJ, objective. e) ii, Schematic of the tip structure for SPP nanofocusing which is composed of a SiO_2 pyramidal structure (AFM tip) and a Ag film on the surface. The incident light is coupled to the surface by the FIB-fabricated grating nanostructure. e) iii, SEM image of a Ag-coated SiO_2 tip with a FIB-fabricated grating structure.

The surface of AFM tips becomes nanostructured during the coating process resembling aggregated nanoparticles on the semiconducting tip. These nanostructures are suitable for the resonant excitation of LSPR and SPPs. The smooth tip shown in figure has been fabricated by subsequently depositing a thin granular layer of additional metal. Other researchers have tested AFM tips with a metallic nanoparticle attached to the tip apex, or segregation in the tips coating to form an antenna by focused ion beam lithography. Tips can also be created by electrochemical deposition.

For transparent dielectric substrates, a thin metal film (thin enough to be transparent) can be coated onto the substrate to further enhance the field in the tip-sample gap. It is also possible to perform TERS in liquids with AFM-based systems, which is favourable for biological specimens which require liquid environments to function. Performing TERS in liquid with STM systems is much more difficult. SFM-based TERS is also an attractive technique and maintains many of the properties of AFM-based TERS with the exception of the tip material which resembles similar TERS properties of STM-based TERS.

Some TERS setups have demonstrated vastly improved signal-to-noise ratio in TERS by SPP nanofocusing. This technique focuses the laser onto a plasmon-coupling nanostructure (in the form of a grating) on the upper area of the tip, usually at a distance of ~10 μm from the tip apex. Figure shows a typical nanofocused SPP-based TERS setup (i), the process of SPP nanofocusing by coupling the incident light to a focused ion beam-fabricated grating (ii), and an example SEM image of a SPP-nanofocusing tip (iii). The excited plasmons then propagate toward the tip apex through the process of adiabatic compression and create a confined EM field at the tip apex.

Tuning the Plasmon Resonance

The size, shape, composition of the nanostructures and inter-nanostructure spacing all affect the wavelength of the surface plasmon resonance. Metals are most often used as the conducting medium for surface plasmons; however, semiconductors also possess plasmonic characteristics.

Gold shows strong enhancement factors in the red spectral region, silver in the blue-green spectral region and aluminium in the UV and deep UV spectral regions. The blue-green spectral region is the most commonly used Raman spectroscopy range. However, silver is prone to oxidation which degrades the plasmonic characteristics within a few hours of exposure to atmosphere. For this reason, silver is often mixed with other metals, such as titanium.

The range of plasmon resonance can be tuned by the thickness and choice of coating metal, e.g. tungsten, gold, silver or aluminium. In TERS, the grain size of the metal coating corrugations is roughly comparable to the wavelength of the LSPR/SSP. Unlike STM tips, it is possible to control the LSPR/SPP wavelength by adjusting the size of the nanoparticles. The surface plasmon resonance wavelength is also dependent on the refractive index of the dielectric material. In AFM-based TERS, for example, the silicon cantilever tip can be heated to ~1000 °C in the presence of water vapour to oxidise the silicon into silicon dioxide. As SiO_2 has a lower refractive index than Si, the surface plasmon resonance is blue shifted.

The size and shape of the metal-coated AFM tip apex can also be modified to tune the LSPR. Fabricating a single metallic nanoparticle attached to the tip's apex or segregated antenna-shaped tip has been demonstrated as a means to finely tune the surface plasmon resonance in AFM-based TERS. However, the most commonly used tips for AFM-based TERS are the tips that have disconnected metal nanoparticles evaporated on a semiconductor cantilever in the standard coating process described a priori.

Analysis Methods

A Note on Units

By convention, Raman spectra are considered in terms of the wavenumber $\tilde{\nu}$ in units of cm^{-1}. The conversion from angular frequency is as follows:

$$\tilde{\nu} = \frac{\omega}{2\pi c_0},$$

where c_0 is the speed of light in vacuum and ω is the angular frequency. Raman spectra are usually plotted in terms of the wavenumber shift from the incident excitation radiation. This shift is defined as follows:

$$\Delta\tilde{\nu} = \tilde{\nu}_{\mathrm{p}} - \tilde{\nu}_{\mathrm{scat}},$$

where $\tilde{\nu}_{\mathrm{p}}$ is the wavenumber of the pump beam with angular frequency ω_p and $\tilde{\nu}_{\mathrm{scat}}$ is the wavenumber of the scattered light accordingly. For Stokes Raman scattering, $\tilde{\nu}_{scat} = \tilde{\nu}_p - \tilde{\nu}_{osc}$ (where $\tilde{\nu}_{osc}$ is the molecule or lattice vibration wavenumber) and $\Delta\tilde{\nu}$ is positive. By contrast, for anti-Stokes Raman scattering, $\tilde{\nu}_{scat} = \tilde{\nu}_p + \tilde{\nu}_{osc}$ and $\Delta\tilde{\nu}$ is negative.

Raman spectra are (by standard) presented with the wavenumber shift linearly increasing from right to left on the horizontal axis. The vertical axis ordinate is linear and proportional to intensity. However, researchers also present Raman spectra with wavenumber shift denoted simply as wavenumber and/or increasing from left to right instead of right to left.

Spontaneous Raman Spectra

Figure shows the Rayleigh and the Raman spectrum of carbon tetrachloride (liquid) excited by an argon ion laser, $\tilde{\nu}_1 \sim 20{,}487\ \text{cm}^{-1}$ (487.99 nm). This spectrum is presented according to recommendations of the International Union of Pure and Applied Chemistry. It contains a strong band at $\tilde{\nu}_1 \sim 20{,}487\ \text{cm}^{-1}$ due to the Rayleigh scattering of the incident laser radiation and a number of weaker bands with wavenumbers, $\tilde{\nu}_1 \pm \tilde{\nu}_{osc} : \tilde{\nu}_{osc} = 218, 314, 459, 762$ and $790\ \text{cm}^{-1}$. The Stokes Raman lines are shown on the left-hand side of the plot; the anti-Stokes Raman lines are shown on the right. The $\tilde{\nu}_{osc}$ values relate to the fundamental vibrations of the carbon tetrachloride molecule. In the original work by Raman and Krishnan, the same spectrum was measured using mercury arc radiation ($\tilde{\nu}_1 = 22{,}938\ \text{cm}^{-1}$, 435.83 nm). In this seminal work, the anti-Stokes bands at $\tilde{\nu}_1 + 762$ and $\tilde{\nu}_1 + 790\ \text{cm}^{-1}$ were not observed. Hence, after the invention of the laser, Rayleigh and Raman scattering experiments are preferably performed using monochromatically intense lasers.

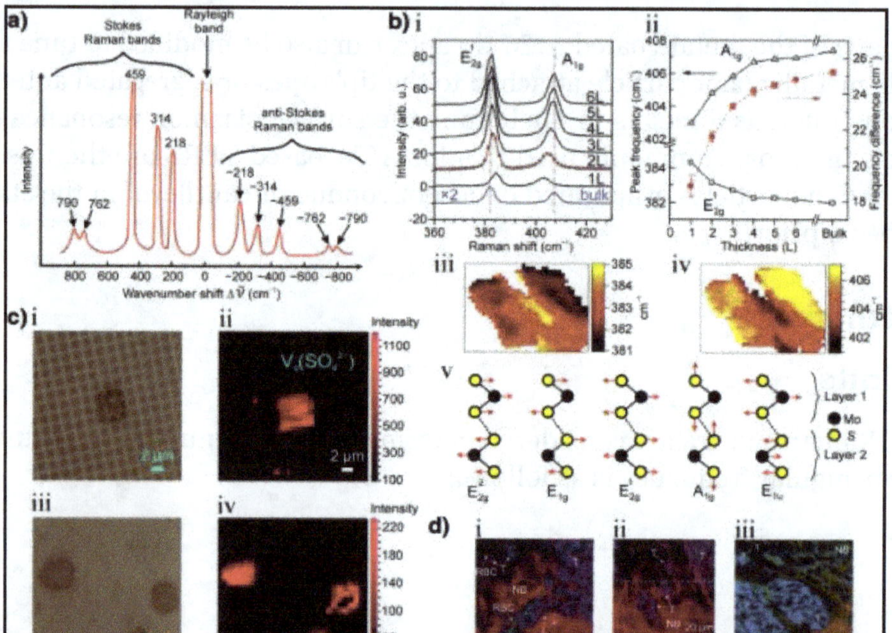

In figure, a) Spontaneous Stokes and anti-Stokes Raman spectrum of carbon tetrachloride (liquid) excited by an argon ion laser, $\tilde{\nu}_p = 20487\ \text{cm}^{-1}$. The spectrum is presented according to recommendations of the International Union of Pure and Applied Chemistry. b) i, Raman spectra of thin multi-layer (nL) and bulk MoS2 films. The solid line for the 2 L spectrum is a double Voigt fit through data (circles for 2L, solid lines for the remainder). b) ii, Frequencies of E^1_{2g} and A_{1g} Raman modes (left vertical axis) and their difference (right vertical axis) as a function of the number of layers. b) iii, iv spatial maps (23 µm × 10 µm) of Raman frequency of E^1_{2g} (iii) and A_{1g} (iv) from a sample of thin MoS2 films deposited on a SiO_2/Si substrate. b) v, Atomic displacements of the four Raman-active modes and one infrared-active mode (E_{1u}) in the unit cell of the bulk MoS2 crystal as viewed along the [1000] direction. c) Microscopic image of nebulised ammonium sulphate aerosol particles on: i, Klarite; iii, silicon wafer. ii, iv Raman mapping image of sample (i) and (iii), respectively. d) i, Pseudo colour broadband CARS image of tumour and normal brain tissue, with nuclei highlighted in blue, lipid content in red and red blood cells in green. d) ii, Broadband CARS image and axial scan (below) with nuclei highlighted in blue and lipid content in red. d) iii, Broadband CARS image with nuclei highlighted in blue, lipid content in red and CH3 stretch–CH2 stretch in green. NB, normal brain; T, tumour cells; RBC, red blood cells; L, lipid bodies; WM, white matter. d) iv, Single-pixel spectra. e) Raman thermography measurements across the active region of a high electron mobility transistor on SiC substrate with both E_2 and A_1 (LO) phonons considered to compensate for thermal stress.

Device temperature rise determined using either E_2 or A_1 (LO) phonon mode alone (neglecting thermal stress) is shown in the top left insert. f) (left) illustration of the manipulation of a straight isolated carbon nanotube (CNT) lying on a glass substrate by the sharp apex of an AFM tip. f) (right) two-dimensional image of a CNT constructed by colour-coding the frequency position of the G+ vibrational mode in TERS spectra. The colour variation shows the strain distribution along the CNT at high-spatial resolution.

Layered Two-Dimensional Systems

Raman spectroscopy can be used to determine the layer thickness in two-dimensional materials with atomic level precision, using either the inter-layer or intra-layer vibrational modes. Lee et al. demonstrated the technique with two intra-layer Raman modes of molybdenum disulphide (MoS_2). Figure shows representative Raman spectra for single- and few-layer MoS_2 samples. Among the four Raman-active modes of bulk 2H phase MoS_2 crystal Lee et al. only observed the E^1_{2g} and A_{1g} modes near $\Delta\tilde{\nu} = 400$ cm⁻¹. The other modes were not observed either because of the selection rules for the scattering geometry $\left(E_{1g}\right)$ or because of the limited rejection of the Rayleigh scattering radiation $\left(E^2_{2g}\right)$. Single-layer MoS_2 exhibits a strong in-plane vibrational mode at $\text{Ä}\tilde{\nu} \sim 384\,\text{cm}^{-1}$, corresponding to the E^1_{2g} mode of the bulk 2H-MoS_2 crystal. For all film thickness, the Raman spectra in Figure (b) (i) show strong in-plane E^1_{2g} and out-of-plane A_{1g} vibration signals. As the sample thickness increases, the E^1_{2g} mode red shifts and the A_{1g} mode blue shifts. For films of four of more layers, the E^1_{2g} and A_{1g} modes converge on the bulk values. Spatial maps of a MoS_2 film sample for the E^1_{2g} mode is shown in Fig. (b) (iii) that of the A_{1g} mode is shown in Fig. (b) (iv). These maps demonstrate that the frequency of the two modes only slightly vary in regions of the sample with a given layer thickness. Hence, Raman spectra can provide a convenient and reliable means of determining the layer thickness in two-dimensional crystalline materials with atomic level precision.

Enhanced Raman Scattering through SERS

Ault et al. were the first to use SERS to enhance the Raman scattering signal of previously undetectable secondary organic aerosol particles on Ag nanoparticle-coated quartz substrates. Fu et al. demonstrated enhancement factors of 6 for the Raman spectra of ammonium sulphate (AS) at the Raman active mode $\tilde{\nu}_s(SO_4^{2-})$ at 970 cm−1 with Klarite. Figure (c) shows a microscope image of a large AS particle on the surface of Klarite, the corresponding Raman mapping image is in Figure. (c) (ii). Figure (c) (iii) shows another sample of AS particle but on a silicon wafer. The corresponding Raman mapping image is shown in Figure (c) (iv). Aside from the three larger AS particles, small (sub-micron) AS particles are apparent in Figure (c) (iii) However, in the absence of SERS, these smaller particles are undetectable. On the other hand, the SERS Raman mapping image shows vastly enhanced signal intensity, as is evident from the scale bars, to the point where a number of small spots yield a signal at the $\tilde{\nu}_s(SO_4^{2-})$ Raman mode. Such spots most likely correspond to small AS particles that are observable in Figure (c) (ii) but are not apparent in Figure (c) (i).

Insights into Cellular Structure with CARS

CARS microscopy is relevant to the chemical, materials, biological and medical fields and can provide unparalleled insights into cellular structures. Spontaneous Raman and infrared micro/spectroscopy can provide adequate chemical specificity and sensitivity to delineate a variety of neoplasms but require long integration times and have a coarse spatial resolution, which may limit accurate tumour-boundary identification and early-stage tumour detection.

However, coherent Raman imaging techniques have demonstrated high-speed, high-spatial-resolution imaging, but with contrast limited to single or few Raman peaks. Figure (d) presents images of orthotopic xenograft brain tumours from within a murine brain. Figure (d) (i) shows a broadband CARS image with nuclei in blue ($\Delta\tilde{\nu} = 730$ cm^{-1}), lipid content in red (Δ $= 2850$ cm^{-1}) and red blood cells in green ($\ddot{A}\tilde{\nu} = 1548$ cm^{-1} + 1565 cm^{-1}: C-C stretch from haemoglobin). The large tumour mass and a projection of neoplastic cells within healthy tissue are clearly shown. Figure (ii) shows several small regions of main tumour mass migrating into the healthy brain matter. Figure (iii) shows the boundary between normal brain tissue, white matter and tumour masses, which contrasts lipids in red ($\Delta\tilde{\nu} = 2850$ cm^{-1}); CH_3 stretch-CH_2 stretch ($\Delta\tilde{\nu} = 2944 - 2850$ cm^{-1}), a general contrast; and nuclei in blue ($\Delta\tilde{\nu} = 785$ cm^{-1}). The image shows the fibrous texture of the white matter and strands of myelination around cancer cell clusters. Figure (d) (iv) presents a set of single-pixel spectra from an intra-tumoural nucleus, the white matter and normal brain, respectively. The spectra indicate that lipids are most concentrated in the white matter and least in the tumour regions.

Raman Thermography

Advances in electronic and opto-electronic semiconductor devices, such as high electron mobility transistors (HEMTs), have led to thermal management challenges. Conventional thermal characterisation approaches such as infrared thermography are often no longer applicable for the accurate characterisation of high-power density devices due to limited spatial resolution which can result in the underestimation of the device peak temperature. Batten et al. have demonstrated temperature profiling in AlGaN/GaN HEMTs using Raman thermography by exploiting the E_2 and A_1 (LO) phonon modes. Both the E_2 and A_1 (LO) modes shift to lower frequency when operating the device. Figure (e) shows a comparison of the temperature rise in a AlGaN/GaN HEMT on a SiC substrate from Raman thermography and thermal simulations. The device was operated at a source-drain voltage of 40 V and a power density of 25 W/mm and had a thermal resistance of 8 °C/(W/mm).

Measuring Strain on the Nanoscale using TERS

TERS microscopy is an effective means of imaging nanostructures beyond the spatial resolution of the so-called light diffraction limit. Nanostructures such as DNA molecules, carbon nanotubes (CNTs), silicon devices, dye molecules and single molecules can be imaged using TERS. The technique can even be used to measure the local molecular strain in nanostructured materials. For example, AFM can be used to manipulate CNTs with nanoscale precision to develop a local strain. Figure (f) (left) illustrates the process of CNT manipulation using contact-mode AFM. Although local strain in CNTs has previously been studied using AFM and transmission electron microscopy, TERS microscopy is the only optical technique that can provide images of such local structural distribution of nanomaterials. When a straight CNT is deformed by manipulation, a local breakdown in symmetry is induced. This causes the selections rules of Raman scattering to become relaxed, allowing forbidden Raman modes to become visible in the vicinity of the local curvature. The position of the characteristic G-mode Raman scattering line in graphene can be used to deduce local strain using TERS. Figure (f) (right) shows a TERS image of a deformed CNT which has been constructed from the peak positions of the G^+-mode. The image has a spatial resolution better than 20 nm which is about 25 times finer than the diffraction limit of the excitation wavelength of light (488 nm). The colour variation (as indicated by the scale bar) corresponds to the local peak position of the G^+-mode and represents the variation of strain along the CNT.

Stimulated Raman Scattering Microscopy

Unlike CARS, SRS microscopy does not contain a nonresonant background signal that degrades image contrast. However, SRS can be affected by cross-phase modulation (where light at one wavelength modulates the refractive index in the medium affecting another wavelength of light), transient-absorption (which is characteristic of femtosecond light pulses) and photo-thermal effects which can modify the vibrational energy levels and reduce hyperspectral image contrast. SRS is quantified by the amount of energy transfer from the pump light to the Stokes light when the difference frequency between the pump and Stokes light matches a specific vibrational frequency, ω_{osc}. In addition, the resulting signal from SRS is strongly sensitive to the incident polarisations when the orientation of the probed vibrating species is ordered. This polarisation dependence can be exploited to probe the orientational order of vibrational modes in samples. However, currently developed techniques are not able to perform large-field fast time scale dynamics instantaneously due to the requirement of point-wise scanning over the sample space. Conventional polarisation-resolved techniques take minutes because each point of the scanning area must be polarisation tuned sequentially.

Multi-lamellar myelin plays a crucial role for efficient transmission of nerve impulses as an electrical insulator. The lipids and proteins in myelin self-assemble into a highly ordered and stable structure to form a tightly packed membrane. In neurological disorders, this compact structure is highly perturbed leading to dysfunctions of the central nervous system. As these biological processes are highly dynamic, researchers seek to observe the dynamics of molecular order with sufficient resolution and frame rate. Hofer et al. have recently demonstrated fast-polar-SRS by exploiting high-speed amplitude- and polarisation modulation with an acousto-optic modulator (AOM) and electro-optical polarisation modulation, respectively, to read out the molecular order and orientation at a fast rate. They therefore obtain both amplitude and phase information. The linear polarisation direction of the pump beam is rapidly rotated while the Stokes polarisation is circularly polarised to avoid polarisation dependence from the Stokes beam. The polarisation is further modified by a quarter-wave plate. The polarisation modulation leads to an α-dependant response of the signal intensity given by the following:

$$I(\alpha) \propto a_0 + S_2 \cos 2(\alpha - \varphi_2)$$

where α is the rotating pump polarisation direction in the sample plane, a_0 is the total measured intensity, and S_2 and φ_2 are the amplitude and phase of the second-order induced modulation.

Figure (a) (i) below shows a comparison of conventional polarisation SRS with that from Hofer's fast-polarisation SRS on a multi-lamellar lipid vesicle (MLV). The fast-polarisation SRS image in the bottom of Figure (a) (i), was obtained in 1 s which is two orders of magnitude faster than the conventional-SRS image (top) using the same incident powers, number of pixels and dwell time per pixel. Figure (a) (ii), shows sub-second frame-rate imaging of a MLV using double EOM-AOM modulation SRS at two instances in time. The measurement technique was remarked to have little effect on the lipid order properties during the measurement. Hofer et al. were able to observe second-timescale dynamics in *thin* lipid membranes down to the cell plasma membrane using fast-polarisation-resolved SRS as shown in figure (a) (iii).

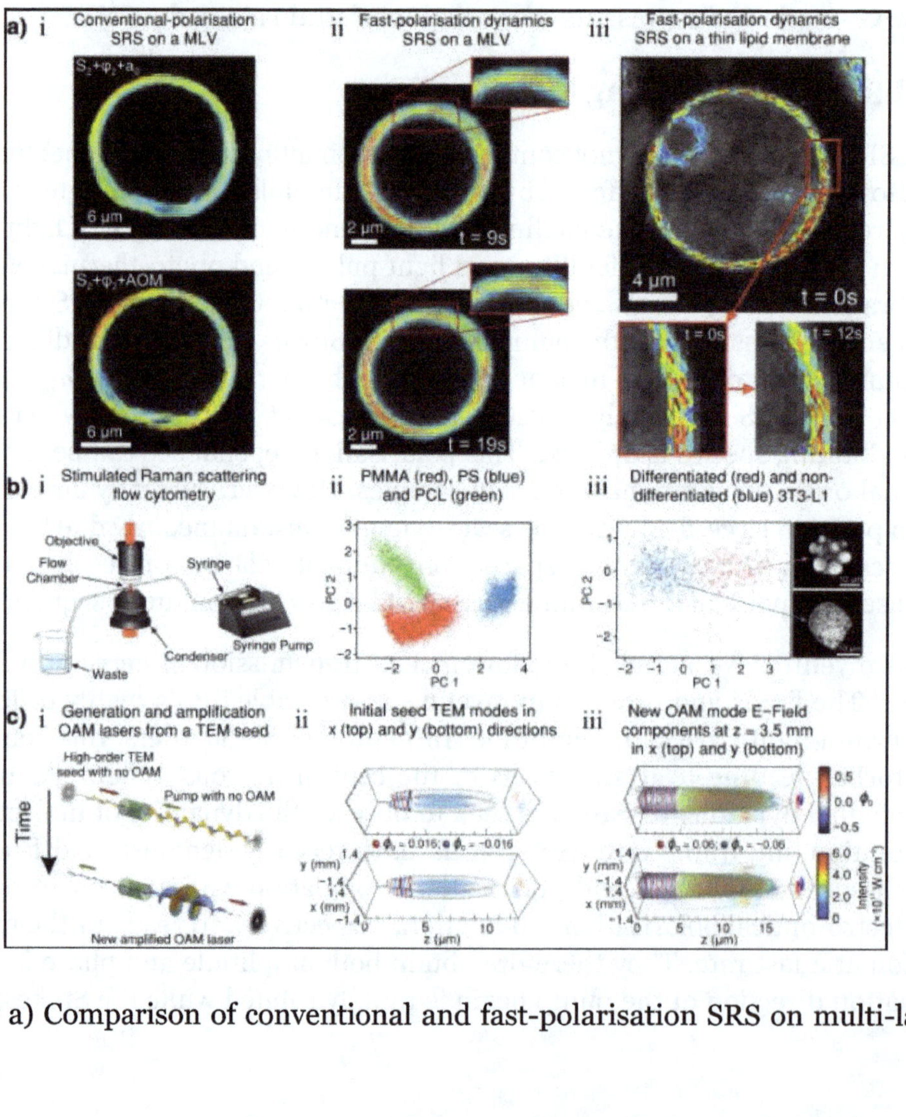

In figure show a) Comparison of conventional and fast-polarisation SRS on multi-lamellar lipid vesicles (MLVs). a) i–iii, the composite images show absolute local molecular order (S_2) and mean molecular orientation (φ_2) values represented as coloured sticks. a) i, (top) Conventional polarisation SRS on a MLV using step angles of 5° (acquisition time 112 s); the total measured intensity (a_0) is represented as a grey-scaled background. a) i, (bottom), fast-polarisation SRS on the same MLV (acquisition time 1 s); the acousto-optic modulation (AOM) is shown as a grey-scaled background. a) ii, Fast-polarisation dynamics SRS images of lipid order in a MLV taken at different times of the observation sequence shown; the AOM is shown as a grey-scaled background. Zoomed regions at the upper part of the MLV contour show no change in lipid order during the measurement over tens of seconds. a) iii, Fast-polarisation dynamics of lipid order in a thin lipid membrane. Coloured sticks show S_2 and φ_2 and the grey-scale background shows the AOM amplitude. b) i, SRS flow cytometry (SRS-FC) setup. b) ii, Colour-coded constrained principle component analysis (CPCA) scatter plot of SRS-FC spectra from mixed PMMA (red), PS (blue) and PCL (green) beads.

The principle components (PC 1 and PC 2) are distinguished subpopulations of mixed polymer beads according to the distinct Raman spectra. Data were acquired in 6 s using a bead mixture with a concentration of 2% solids. The beads were 10 µm in diameter. b) iii, (left) Colour-coded CPCA scatter plot from SRS-FC analysis of lipid amount in 3T3-L1 cells with principle components differentiated through quantification of distinct chemical compositions inside single cells. Data were acquired in 3 s. b) iii, (right) SRS images of the two 3T3-L1 cell types. c) i, Illustration of the generation and amplification of a new orbital angular momentum (OAM) laser in a configuration with no initial OAM using a Hermite-Gaussian (TEM) laser. c) ii, iii, Simulation of the generation and amplification of a new OAM mode from initial configurations with no net OAM. c) ii, The initial seed TEM modes in x- and y-directions (top and bottom, respectively). c) iii, The new OAM mode electric field components at $z = 3.5$ mm in x- and y-directions (top and bottom, respectively).

The new mode is linearly polarised in the x- and y-directions with $l_{1x} = l_{1y} = -1$ from an initial seed polarised in the x-direction with a TEM01 mode and in the y-direction with a TEM10 mode that is $\pi / 2$ out of phase with respect to the TEM01 mode polarised in x. The pump is a Gaussian laser polarised at 45°. Projections in the (x, y) plane (blue-white-red) show the normalised vector potential $(\phi 0)$ field envelope of the new OAM mode at the longitudinal slice where the laser intensity is maximum. The envelope of the 3D laser intensity is also shown in blue-green-red colours and normalised vector potential isosurfaces in blue and red.

Flow cytometry (FC) is one of the most important technologies for high-throughput single-cell analysis. FC is a technique used to measure physical/chemical characteristics of a population of cells or particles suspended in a fluid. The fluid suspension flows through the instrument detectors for fluorescent labelling which is the primary approach for cellular analysis in FC. Figure (b) (i), shows an optical FC setup. However, for small molecules, the fluorescent tags can perturb the biological function of the species. In addition, non-specific binding of fluorescent labels as well as cellular autofluorescence can also reduce the clarity of the result. SRS flow cytometry (SRS-FC) non-invasively detects chemical cell content but conventional techniques suffer slow acquisition rates.

Zhang et al. have recently demonstrated label-free high-throughput single-particle SRS-FC with a 32-channel multiplexing technique. Their technique measured single-particle chemistry at a rate of 5 µs per SRS spectrum, approaching that of standard fluorescence-based FC. The SRS-FC technique was based on broadband laser excitation and a multiplex spectral detection system. The systems allowed the acquisition of 200,000 spectra per second, more than 11,000 particles per second. The subpopulations of species, such as mixed polymer beads and 3T3-L1 cells, could be separated and distinguished through compositional principle component analysis (CPCA) of the SRS signals. The principle components were designated according to their Raman spectra. An agglomerative clustering procedure was performed on the resulting CPCA spectral matrix.

This procedure assumed the number of cluster groups (κ) was known to separate the clusters of principle components in the CPCA analysis. Figure (b) (ii), shows the CPCA of the SRS spectra for a mixture of three types of beads: poly-methyl-methacrylate (PMMA), polystyrene (PS) and polycaprolactone (PCL), all with a 10-µm mean diameter, mixed at a 2:1:1 ratio of PMMA:PS:PCL and a final concentration of 2% solids in the fluid. The flow speed was $\approx 0.16\,\text{ms}^{-1}$, the SRS-FC data was acquired in 6 s. The CPCA plot shows three distinct clusters of principle components. The agglomerative clustering procedure $(\kappa = 3)$ allowed the quantification of ~7100 PMMA bead (red), ~3400 PS beads (blue) and ~3600 PCL beads (green) as shown in figure (b) (ii). Their measurement demonstrated the ratio of ≈2:1:1 (PMMA:PS:PCL) at a throughput rate of 2350 particles per second and that their multiplex SRS-FC system, paired with the CPCA analysis, could distinguish different chemical components with small spectral differences. Zhang et al. were able to detect beads as small as 1 µm and were even able to detect single Staphylococcus aureus bacteria flowing through the laser focus highlighting the potential to characterise subcellular organelles with SRS-FC.

Zhang et al. also demonstrated the discrimination of 3T3-L1 cells at different stages of cell differentiation according to their difference in lipid amount using SRS-FC. After insulin-induced differentiation, 3T3-L1 cells acquire an adipocyte-like phenotype with a significantly increased amount of triglycerides which aggregate to form large lipid droplets. This aggregation of triglycerides causes the intensity of the methylene symmetric vibration at $\text{Ä}\tilde{\nu} = 2850\,\text{cm}^{-1}$ from fatty-acid acyl chains to become stronger compared to that of non-differentiated cells which provides the means for CPCA analysis. Figure (b) (iii) shows the CPCA scatter plot of the cell mixture measured by Zhang et al. which were separated using the agglomerative clustering approach. The insert SRS images show a non-differentiated 3T3-L1 cell and a differentiated cell with the formation of large lipid droplets.

Twisted Laguerre-Gaussian lasers, with orbital angular momentum (OAM) and characterised by doughnut-shaped intensity profiles, are of great interest to a number of growing research fields such as ultra-cold atoms, microscopy and imaging, atomic and nanoparticle manipulation, ultra-fast optical communication, quantum computing, astrophysics and plasma accelerators. Spiral phase plates or computer-generated holograms are usually used to generate visible light with OAM. Spiral phase plates can produce light with predefined OAM modes. By using plasma as an optical medium to generate and amplify laser pulses with OAM and relativistic intensities, well above the damage threshold of optical devices, could provide for high-energy-density science and applications. Plasmas also allow for greater flexibility in the level of OAM in the output laser beam than conventional optics. Vieira et al. have shown that SRS in nonlinear optical media with Kerr nonlinearity (e.g. plasmas, optical fibres and nonlinear optical crystals) can be used to generate and amplify OAM light. Figure (a) (i) illustrates the process in which the pump EM fields can have different OAM components in both transverse directions x and y (blue and orange in Figure (a) (i)). l_{0x} is the pump electric field component of OAM in the x direction. Likewise, l_{oy} is the pump electric field component of OAM in the y direction. The initial seed electric field component has an OAM component l_{1x}. After interacting with the plasma, the pump is depleted, and a new electric field component appears in the seed with OAM $l_{1y} = l_{1x} + l_{0x} - l_{0y}$.

Stimulated Raman backscattering in plasma is a three-wave mode coupling process in which a pump pulse decays into an electrostatic (Langmuir) plasma wave as well as a counter-propagating seed laser. The plasma can be viewed as a high-intensity mode converter. The presence of OAM in the pump and/or seed results in additional matching conditions that ensure the conservation of angular momentum of the pump when the pump decays into a scattered electro-magnetic wave and a Langmuir wave.

Particular superpositions of Hermite-Gaussian modes TEM modes are mathematically equivalent to Laguerre-Gaussian modes. Vieira et al. therefore explore the use of Stimulated Raman backscattering to generate and amplify light with OAM using TEM laser beams with no initial net OAM. Each Hermite-Gaussian beam in the simulation is described by $TEM_{m,n}$, where m and n represent the Hermite-Gaussian mode index. Figure (c) (ii) and iii show the 3D simulation results from the setup shown in Figure (c) (i). The simulations show that SRS results in a new OAM mode with $l_1 = 1$ linearly polarised at 45°. The field topology of the seed normalised vector potential changes from plane isosurfaces in Figure (c) (ii), to helical isosurfaces in. Hence, light with OAM has been generated from light with no net OAM. In the case of plasma, the interaction between the seed light and the pump light occurs via an electron Langmuir wave. This interaction ensures that the frequency, wavenumber and OAM matching conditions are conserved.

Coherent Anti-Stokes Raman Scattering Microscopy

CARS results from an induced anti-Stokes scattering of radiation, ω_{AS}, which is enhanced when $\omega_p - \omega_S = \omega_{OSC}$. One of the main challenges with CARS microscopy is the nonresonant background. The existence of the nonresonant background can either distort or even saturate the resonant signal of Raman peaks, which reduces the image contrast. Qin et al. have recently demonstrated multi-colour background-free coherent anti-Stokes Raman scattering microscopy using an all-fibre, low-cost, multi-wavelength *time lens* source. A time lens, in analogy to a spatial lens, is simply a quadratic optical phase modulator in time, which can be approximated by a portion of a sinusoidal phase modulator. Three different wavelength picosecond pulse trains were provided by the time lens source, at 1064.3 nm (stable), 1052–1055 nm (tuneable) and 1040–1050 nm (tuneable).

The time lens was used to apply temporal quadratic phase modulation to a continuous-wave laser to broaden its spectral bandwidth. In this instance, the time lens was applied with fibre-integrated electro-optic radio-frequency phase modulators. The phase modulation and pulse synchronisation were derived from a mode-locked Ti:Sapphire laser that provided synchronised multi-colour picosecond pulses with dispersion compensation. Electronic tuning of the pulse delay was used to achieve temporal overlap between the pump and Stokes laser pulse trains, which is a convenient substitution for mechanical optical delay paths. Two of the three wavelengths of light from the time lens source were used for two-colour on-resonance imaging and the third wavelength for off-resonance (nonresonant background subtraction) imaging. Pixel-to-pixel wavelength switching was achieved, which provided simultaneous two-colour CARS imaging with real-time nonresonant background subtraction. Qin et al. demonstrated the technique with an excised fresh tissue sample from a mouse ear and imaged molecular stretching vibrations at 2845 cm^{-1} (CH_2) and 2940 cm^{-1} (CH_3) and non-resonance background at $Ä\tilde{\nu} = 2770$ cm^{-1}. Figure (a) (i–iii) shows the process applied to the Raman peak of CH_3 stretching vibration from the mouse ear tissue sample.

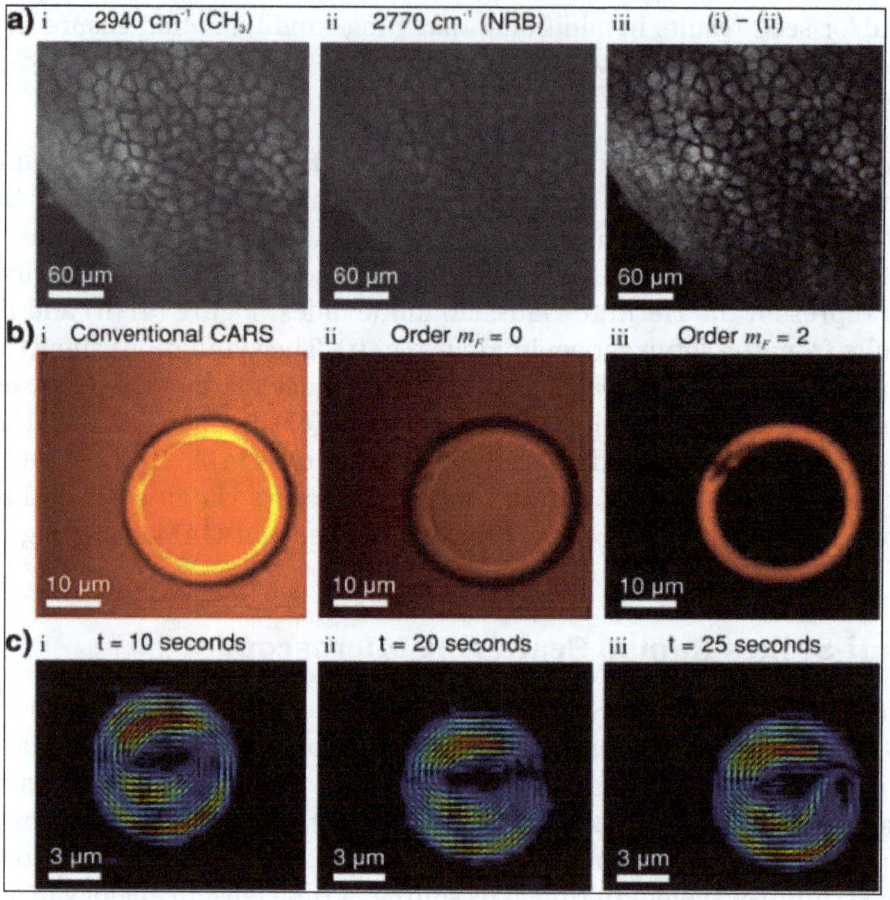

In figure, a) Simultaneous two-colour CARS imaging with real-time nonresonant background subtraction from a mouse tissue sample at a surface depth of 45 µm. a) i, CARS image acquired at 2940 cm^{-1} (CH3 stretching vibration). a) ii, Off-resonance background CARS image at 2770 cm^{-1}. a) iii, Background-free image of (i) at 2940 cm-1 after subtraction of the nonresonant background (ii). b) Multi-lamellar lipid vesicle (MLV) imaged with conventional (i) and symmetry-resolved CARS (SR-CARS) at 1133 cm^{-1} (ii) and (iii). ii The incident circularly polarised pump, Stokes, probe and anti-Stokes light have co-rotating handedness ($m_F = 0$). Iii, the incident circularly polarised pump, probe and anti-Stokes light have co-rotating handedness and the Stokes light has counter rotating handedness ($m_F = 2$). c) High-speed polarisation-resolved CARS image sequence on a MLV moving over the sample surface taken at different times of the observation sequence shown as a composite image of S_2 and φ_2 as coloured sticks and with the acousto-optic modulation (AOM) as a grey background.

The nonresonant background signal in CARS can also be suppressed by applying an external static electric field to the sample known as electro-CARS. Capitaine et al. demonstrate this electro-optical technique on n-alkanes in solution with broadband multiplex coherent anti-Stokes Raman scattering spectroscopy. The nonresonant background is suppressed due to the orientation response of the molecules to the electric field. The molecular orientation is related to the induced electric dipole moment. The enhancement of the CARS signal-to-noise ratio was achieved in the case of the CH_2 and CH_3 symmetric/asymmetric stretching vibrational modes.

Conventional CARS provides information about the chemical nature but not about the molecular organisation or symmetry in the system. The Cartesian components of the nonlinear susceptibility tensor $\chi^{(3)}$ represent the vibrational symmetry properties of the material. These tensor elements can be extracted with polarisation-resolved coherent Raman scattering schemes. However, these schemes often involve the acquisition of multiple images from different polarisation angles requiring long acquisition times due to limits imposed by polarisation tuning and time-consuming post-processing.

Cleff et al. have recently demonstrated a label-free microscopy technique that uses circularly polarised light to probe the symmetry as well as the chemical fingerprint of the probed sample in a single acquisition. This symmetry-resolved CARS (SR-CARS) depends on both the presence of (ro-)vibrational modes as well as their local organisation. By switching between combinations of left- and right-handed circular polarisation states for the involved fields, the individual symmetry contributions of the sample can be imaged. This technique offers a straightforward means to access the local organisation of (ro-)vibrational bonds with improved image contrasts (with 1 to 2 orders of magnitude) for anisotropic samples, as well as improved chemical selectivity without post-processing and independently of sample orientation in the transverse plane. In addition, SR-CARS provides higher chemical selectivity with the contrast in symmetry characteristics, which are not accessible with conventional spontaneous Raman or SRS microscopy.

Multi-lamellar lipid vesicles (MLVs) are made of a tight packing of lipid layers forming a ring of highly ordered matter with twofold symmetry and a lipid orientation distribution close to a Gaussian angular shape. Figure (b) (i), above shows a conventional CARS image of an aqueous MLV at $\ddot{A}\tilde{\nu} = 1133$ cm⁻¹ (C-C stretching vibration) which illustrates the expected poor contrast due to the nonresonant background. Figure (b) (ii), and (iii) show the zeroth and second-order $m_{\bar{F}}$ value image of the same MLV as in Figure (b) (ii). The $m_{\bar{F}}$ value is the summation of the light circular polarisation handedness quantum numbers of the incident light beams:

$$m_{\bar{F}} = m_{\mathrm{p}} - m_{\mathrm{s}} + m_{\mathrm{pr}} - m_{\mathrm{as}}$$

When light with field tensor $\bar{F}$ probes matter with nonlinear susceptibility tensor $\chi^{(3)}$, in a CARS process, the light probes only the parts of the matter with identical rotational invariant symmetries (i.e. identical $m_{\bar{F}}$). Hence, by engineering the field tensor of the light, specific sample symmetries can be directly read out, creating a symmetry-based image contrast mechanism. The aqueous solution surrounding the MLV is only visible in the $m_{\bar{F}} = 0$ image due to its purely isotropic nature. Background-free imaging of the MLV with superior contrast with respect to conventional CARS is shown in Figure (b) (iii), at $m_{\bar{F}} = 2$ which results from the symmetric microscopic organisation of the lipids in the MLV. Imaging at $m_{\bar{F}} = 4$ (not shown) lacked sufficient signal strength to provide an image of the MLV due to the lack of anti-symmetry in the lipid organisation.

As with SRS, the CARS signal is sensitive to the polarisation of the incident light when the orientation of the scattering species is ordered. Polarisation-resolved CARS (PR-CARS) requires monitoring of the CARS signal response depending on the relative rotation of the incident light polarisations (pump and Stokes) to the sample, in species with ordered orientations. Provided that the molecular bonds are oriented, the detected intensity of the anti-Stokes signal is maximised when the incident polarisations lie along the averaged direction of the bonds. The ability to monitor lipid order without the need for fluorescent labels can provide information on lipid packing properties. PR-CARS schemes often involve long acquisition times due to limits imposed by polarisation tuning and time-consuming post-processing.

In addition to fast-polar-SRS, Hofer et al. have recently demonstrated fast-polar-CARS imaging with combined electro-optic polarisation and acousto-optic amplitude modulations. Figure (c) shows fast-polarisation CARS with similar sensitivity to that of SRS. Despite the requirement of lock-in amplification for the detection of low modulation over a large nonresonant background, the fast-polarisation technique demonstrated by Hofer et al. can considerably improve the signal-to-noise ratio in CARS imaging. Despite the robustness of MLVs, occasional alteration of molecular order in MLVs could be observed at the time scale accessible in Hofer's experiment (0.25–1 s per image). MLVs could detach from the sample surface, inducing motion or shape change. The modifications observed in Figure (c) were attributed to a local membrane disruption, followed by its spontaneous reformation. Hofer et al. demonstrated the possibility of visualising local modification during MLV displacement that was not accessible using the minute-time-scale conventional polarisation Raman experiments.

There have also been a number of developments in CARS flow cytometry (CARS-FC). However, these techniques were shown to be much slower than fluorescence-based FC. Out-of-focus microparticles can randomly impede CARS-FC and the fluid often generates a strong nonresonant background limiting CARS-FC from achieving high-throughput single-cell analysis. Recently, however, O'Dwyer et al. have demonstrated that it is possible to significantly enhance the fraction of unambiguously and instantly recognised in-focus microparticles, in unconstrained flows by co-monitoring CARS-FC with linear scattering of light.

CARS is invariably performed with two synchronised picosecond laser sources owing to the coherence life time of Raman resonance. Ti:Sapphire oscillators or optical parametric oscillators pumped by a picosecond frequency-doubled Nd:Vanadate laser are the instruments of choice, which are generally very expensive and the synchronisation mechanisms can be challenging. In addition, the spectral drift in the pump wavelength can introduce errors in the calculation of ω_{osc}. Langbein et al. have demonstrated CARS micro-spectroscopy using a single Ti:Sapphire laser oscillator and simple passive optical elements. Vibrational excitation, tuneable over a large spectral range with adjustable spectral resolution, was achieved by spectral selection with dichroic mirrors and linear chirping by glass elements.

Tip-Enhanced Dual Wavelength Coherent Anti-Stokes Raman Scattering Microscopy

TERS offers spatial resolution far beyond the diffraction limit of the probing light. The more conventional technique is to directly illuminate the tip-sample cavity. This technique achieves the desired resolution (beyond the diffraction limit) by forcing the evanescent light into the far field image formation. However, the far field light presence in the tip-sample cavity generates an unfavourable background light source. It is possible to perform TERS by coupling the far-field excitation light to the tip a few tens of micrometres from the tip apex. Femtosecond laser pulses can be coupled to the tip surface by shining the light on a grating fabricated on the tip surface. The SPPs then propagate to the tip apex and generate background-free localised optical excitation.

Toma et al. previously demonstrated selective excitation of a single Raman mode and its CARS imaging of CNT using ultra-fast SPP pulse nanofocusing using an Au tapered tip. In a more recent publication, seminal work by Tomita et al. demonstrated simultaneous nanofocusing of ultra-fast SPP pulses at 440 and 800 nm, which were coupled with a common diffraction grating structure. Figure (a) (i), below illustrates the scheme. The Al-tapered tip had an apex radius of ≈ 35 nm. Selective CARS microscopy that combined an 800 nm (ω) SPP pump pulse and a 440-nm (2ω) SPP probe pulse was achieved. Figure (a) (ii) illustrates the energy level process of $\omega-$ and 2ω-CARS. The pump pulse achieves selective vibrational excitation by spectral focusing. Raman shift intensities with this 2ω-CARS scheme were reported to increase by as much as 4 compared with that of ω-CARS for monolayer graphene.

The selectivity of vibration band excitation and background noise suppression were confirmed on the CARS intensity probed by a 2ω-SPP plasmon pulse for a monolayer graphene sample. Venezuela et al. reported the Raman lines in graphene associated with both phonon-defect processes (such as the D line at $Ä\tilde{\nu} \sim 1350$ cm^{-1}) and two-photon processes (such as the 2D line). The 2D-band intensity in graphene was reduced monotonously when the defect concentration was increased, contrary to the D-band. Tomita et al. applied their multi-vibrational-mode 2ω-CARS imaging method to a multi-walled CNT (MWCNT) at the D, G and 2D bands. This dual-wavelength nanofocusing technique could open new nanoscale micro-spectroscopy and optical excitation schemes in SPM, such as sum frequency mixing, two-photon excitation $(\omega + 2\omega)$ and pump-probe schemes.

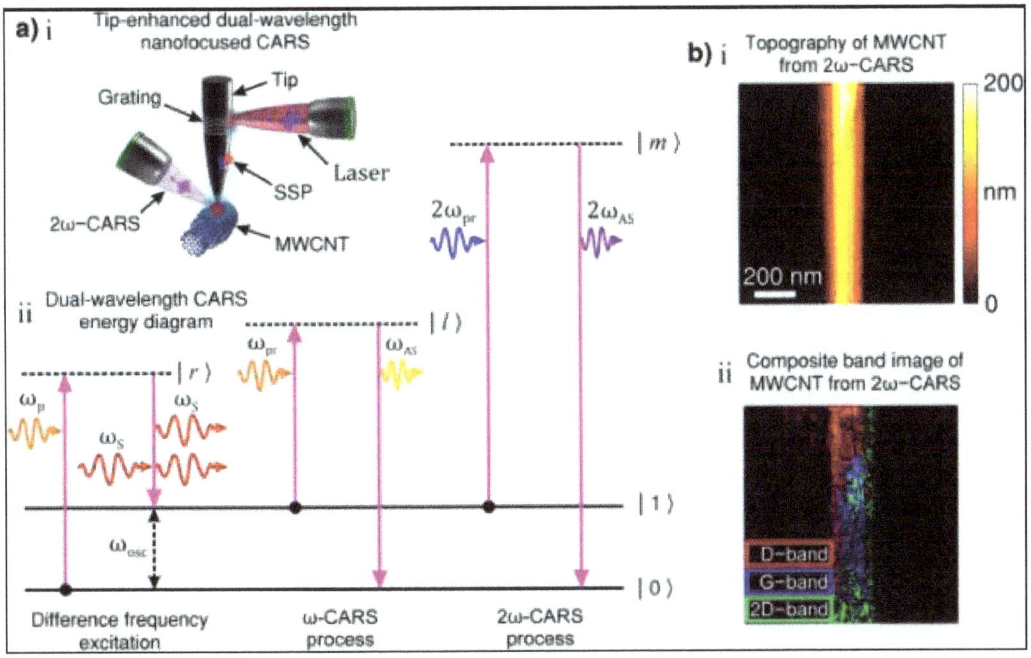

In figure, a) i, Illustration of tip-enhanced dual-wavelength nanofocused CARS on a multi-walled carbon nanotube (MWCNT). a) ii, Energy diagram for ω-CARS and 2ω-CARS. When the difference frequency between pump and Stokes light matches the vibrational mode of the molecule, it is resonantly excited. When ω and 2ω-probe light is simultaneously incident within the dephasing time, ω- and 2ω-CARS photons are respectively generated. b) i, Simultaneous topographical 2ω-CARS imaging of a MWCNT. b) ii, Composite image of three 2ω-CARS images of the MWCNT using the 2ω-CARS spectrum from D- (red), G- (blue) and 2D- (green) bands.

Figure (b) (i), shows a topography image of a MWCNT with a diameter of ~ 175 nm measured by the Al-tapered tip. Figure (b) (ii), shows a composite image of three 2ω-CARS images of the MWCNT using the 2ω-CARS spectrum from D-(red), G-(blue), and 2D-(green) bands. The D- and 2D-band showed a negative correlation except for the central part of the MWCNT. The 2D- and G-bands were intense near the central part of the MWCNT. Tomita et al. indirectly estimated the spatial resolution of their technique to be less than 90 nm by taking the profile of the 2D-band signal across the axis of the MWCNT.

References

- Spectroscopy, cosmos: swin.edu.au, Retrieved 23, January 2020
- Mass-spectrometry, chemistry: byjus.com, Retrieved 14, April 2020

4

Thermoanalytical Methods

Thermal analysis is a technique used to analyze the materials on the basis of their change with temperature. Some of the thermoanalytical methods include differential thermal analysis, differential scanning calorimetry and thermogravimetric methods. The chapter closely examines these methods of thermal analysis to provide an extensive understanding of the subject.

When matter is heated it undergoes certain physical and chemical changes. These changes which are physical or chemical in nature occur at widely varying temperatures. For example, changes such as melting or boiling which are physical in nature may take place over a wide temperature range, and the temperature range is governed by the material involved in the analysis. Changes which are chemical in nature, such as decomposition, may also occur at varying temperatures.

The changes either physical or chemical in nature which take place when a sample undergoes heating are characteristic of a substance under examination. By measurement of the temperature at which such reactions take place and the heat involved in the reaction, the substances present in the material can be easily characterized. The majority of known inorganic compounds have been so characterized. If heating of an unknown sample is done than the physical and chemical changes taking place in the reaction help us to provide the information regarding the identity of the material under investigation. These changes which take place during the reaction give an indication of the temperature at which the material under examination stops to be stable under normal conditions. This information is very useful to industrial chemists, such as those who make varnishes and paints, since it allows them to predict the service lifetime of such compounds. In recent times, thermogravimetric methods of analysis have become very popular in analytical chemistry. The term thermal analysis incorporates those methods in which determination or recording of system's physical parameters is done as a function of temperature.

Table: The various techniques of thermal analysis.

S. No.	Technique	Abbreviations	Instrument used	Parameter measured	Graph obtained
1.	Thermogravimetry	TG	Thermobalance	Mass	Mass vs. Temperature or Time
2.	Derivative Thermogravimetry	DTG	Thermobalance	dm/dt	dm/dt vs. Temperature
3.	Differential Thermal Analysis	DTA	DTA apparatus	ΔT	ΔT vs. Temperature
4.	Differential Scanning Calorimetry	DSC	Calorimeter	dH. dt	dH. dt vs. Temperature
5.	Thermometric Titrimetry	--	Calorimeter	Temperature	Temperature vs. Titrant volume
6.	Dynamic Reflectance Spectroscopy	DRS	Spectrophotometer	Reflectance	%Reflectance and Temperature

7.	Evolved Gas Detection	EGD	Thermal Conductivity Cell	Thermal Conductivity	T.C. vs. Temperature
8.	Thermomechanical Analysis (Dilatometry)	TMA	Dilatometer	Volume of Length	Volume or Length vs. Temperature
9.	Electrical Conductivity	EC	Electrometer or Bridget	Current (I) or Resistance (R)	I or R vs. Temperature
10.	Emanation Thermal Analysis	ETA	ETA apparatus	Radioactivity (E)	E vs. Temperature

Thermal changes are generally studied by recording changes taking place while the temperature is varied which results in a thermal analysis curve or thermogram. It is important to mention here that single thermal analysis method does not provide complete information of a system.

Differential Thermal Analysis

Differential thermal analysis (DTA) in comparison to thermogravimetry is, in fact far more versatile and provides data which is more fundamental in nature. Differential thermal analysis is a very simple technique where the difference in temperature between a material and a reference substance against temperature (or time) is recorded as the two samples are allowed to undergo identical temperature regimes in a controlled environment. Therefore, a differential thermogram is a plot that consists of temperature difference between a sample and reference (differential temperature, ΔT). This difference in temperature is plotted as a function of time (t), temperature of sample (T_s), temperature of reference (T_r) or temperature of furnace (T_f). A typical DTA curve is shown in following figure which consists of two types of peaks; an exothermic peak and an endothermic peak. A peak's shape and size may provide valuable information about the nature of material under investigation.

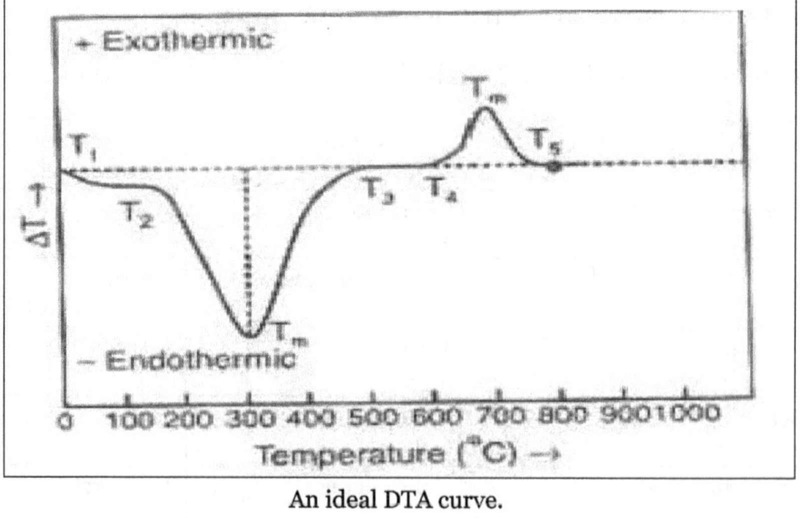

An ideal DTA curve.

Generally, sharp endothermic peaks give ideas of changes in crystallinity or fusion processes whereas board endotherms signify dehydration reactions. Generally, in many cases, changes in physical parameters give rise to curves which are endothermic in nature whereas chemical reactions (particularly those of an oxidative nature) give rise to exothermic peaks.

The temperature difference in the test substance (ΔT) is due to the difference in energy between the products and the reactants of a reaction or between the two phases of a material. This difference in the energy is manifested as enthalpy changes which are either exothermic or endothermic in nature. The DTA curve of a sample would be parallel to either time or temperature axis till any physical and chemical change of state takes place in the material under investigation. The detection of every physical or chemical change can be done easily with DTA.

Factors Affecting DTA Curve

The DTA curve is affected by a larger number of factors than the TG curve. The special influences on DTA curves will be considered. The various factors which may affect a DTA curve are as follows:

Factors Related to Environment

- Sensitivity of DTA technique towards the gaseous environment around the sample is very high in comparison to TG.
- Two types of gaseous atmospheres are employed in differential thermal analysis: (i) static gaseous atmosphere and (ii) dynamic gaseous atmosphere.
- Reproduction of a static gaseous atmosphere is very difficult as the concentration of atmosphere which surrounds the sample changes chemically because of the evolution of various gases. Convection currents are responsible for physical changes during the analysis. Therefore, the studies done in the presence of a static gaseous environment are not precise.
- Dynamic gaseous atmosphere is widely used as the results obtained are reliable and reproducible. The gases, either reactive or inert in nature, produced during the analysis are swept past the sample in a controlled manner.

Factors Related to the Instrument

- Sample holder: The geometry and the material used in the fabrication of the sample holder affect resolution, shape and size of peaks obtained in differential thermal analysis. Materials whose thermal conductivity is very high are used in the construction of sample holders (e.g. metals).
- Differential temperature sensing devices: Generally, heats of transitions are much less as compared to the heats of reactions. If the wires which are being used in temperature sensing devices are much thick, then distortion of the heights of various peaks may be encountered. If a wire which is thin is used in these devices then, a very less amount of distortion in peak heights is observed.
- Characteristics of the furnace: The type of winding used in the furnace has a very profound effect on the curves obtained by differential thermal analysis. A change is observed in the base line if the winding which is used in the furnace is non-uniform. This type of effect is generally seen in hand wound furnaces.

- Temperature programmer controller: There is requirement of temperature controlling during differential thermal analysis as a constant heating rate is necessary to obtain satisfactory results.
- Thermal Regime: Rate of heating employed in the differential thermal analysis has a great effect on DTA curves. If higher heating rates are used then sharper peaks with greater intensity are obtained. Generally, heating rates of about 10 to 20° per minute are used.
- Recorder: Curves obtained by differential thermal analysis are affected by the type, chart-speed, span and pen- response of a recorder.

Sample Characteristics

Physical

Heat capacity remains constant with progress of the reaction. But this is changing with the progress of the reaction. The effect of particle size is related to the effects of density of packing and overall thermal conductivity of material being used as the sample. In general, following interferences are drawn from the variation in size of the particle:

- Size of the particle alters peak area.
- Size of the particle has a great influence on the peak temperature.
- Size of the particle also alters completion temperature.

Chemical

Peaks obtained in differential thermal analysis are greatly affected by chemical reactivity of sample, sample holder, material from which thermocouple is made, ambient gaseous environment and added diluents.

Instrumentation used in DTA

Many different types of instruments are available for DTA studies. DTA apparatus is shown in figure. The various components are as follows:

- Furnace: This device is used for heating sample.
- Sample holder: This is used to contain the sample as well as reference material.
- DC Amplifier: Generally, a low-level DC amplifier is employed.
- Differential temperature detector: The function of this detector is to measure differential temperature.
- Furnace temperature programmer: The main function of this is to increase the temperature of the furnace at a steady rate.
- Recorder: This is to record the DTA curve.
- Control equipment: Its function is to maintain a suitable atmosphere in the furnace and sample holder.

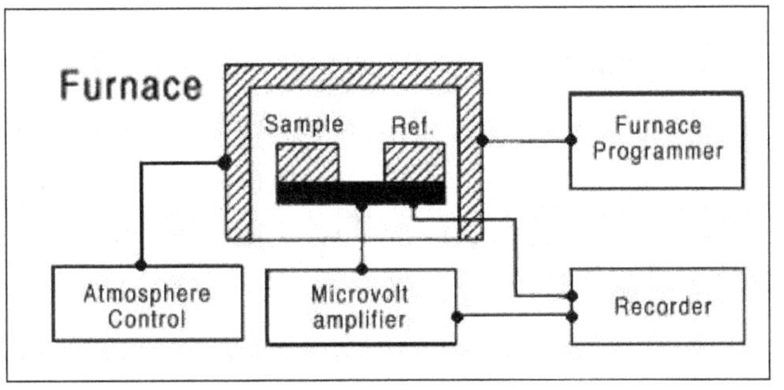

A schematic diagram of a typical DTA apparatus.

Applications of Differential Thermal Analysis

Physical Chemistry

Some main applications are as follows:

- Heat of reaction
- Specific heat
- Thermal diffusivity

Analytical Chemistry

The various applications are as follows:

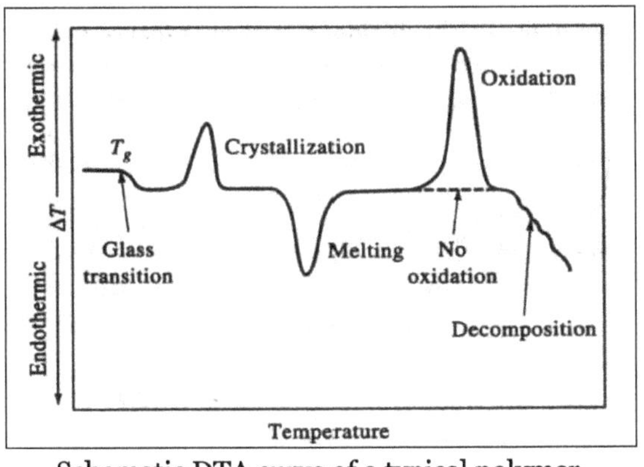

Schematic DTA curve of a typical polymer.

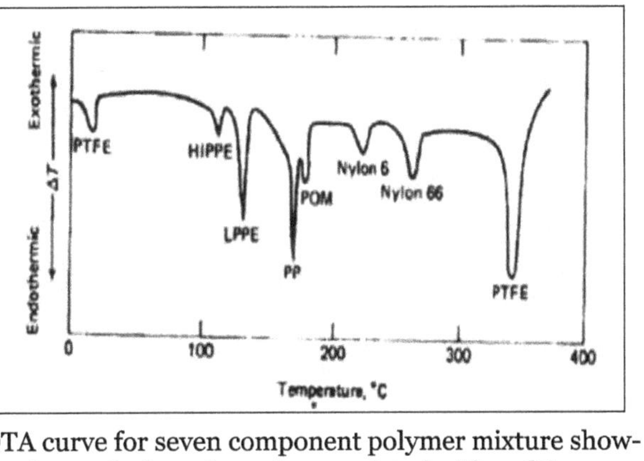

DTA curve for seven component polymer mixture showing endothermic peaks associated with melting.

- Qualitative analysis: Curves obtained by differential thermal analysis for two compounds are not identical. Therefore, this technique can serve as a finger print for different materials.
- Identification of products: Products which are formed when a compound reacts with another compound, can be easily identified by the curves obtained by means of differential thermal analysis.
- Melting Points: As melting point can be easily determined by DTA, it means that this technique can be used as a direct check of the purity of the compound.
- Quantitative Analysis: Areas of DTA peaks are proportional to total heat of reaction and

hence to sample weight. Thus, the amount of a sample can be easily calculated with the help of standard curves of peak area vs. weight.

Inorganic Chemistry

- Differential thermal analysis has been employed for studying thermal stability of different inorganic substances and complexes.
- DTA curves are more helpful in the identification of intermediate compounds.
- Oxalates, carbonates, oxides and metal amine complexes can be easily determined by differential thermal analysis.
- DTA has been used to study the oxides of uranium and plutonium.

Organic Chemistry

Differential thermal analysis has been widely used in identification of substances, determination of purity of different compounds and quantitative analysis including the determination of kinetic parameters of organic chemicals, oils, polymers, explosives, fats and pharmaceuticals.

Differential Scanning Calorimetry

Differential scanning calorimetry is a specific type of calorimetry including both a sample substance and a reference substance, residing in separate chambers. While the reference chamber contains only a solvent (such as water), the sample chamber contains an equal amount of the same solvent in addition to the substance of interest, of which the $\Delta_r H$ is being determined. The $\Delta_r H$ due to the solvent is constant in both chambers, so any difference between the two can be attributed to the presence of the substance of interest. Each chamber is heated by a separate source in a way that their temperatures are always equal. This is accomplished through the use of thermocouples; the temperature of each chamber is constantly monitored and if a temperature difference is detected, then heat will be added to the cooler chamber to compensate for the difference. The heating rate used to maintain equivalent temperatures is logged as a function with respect to the temperature.

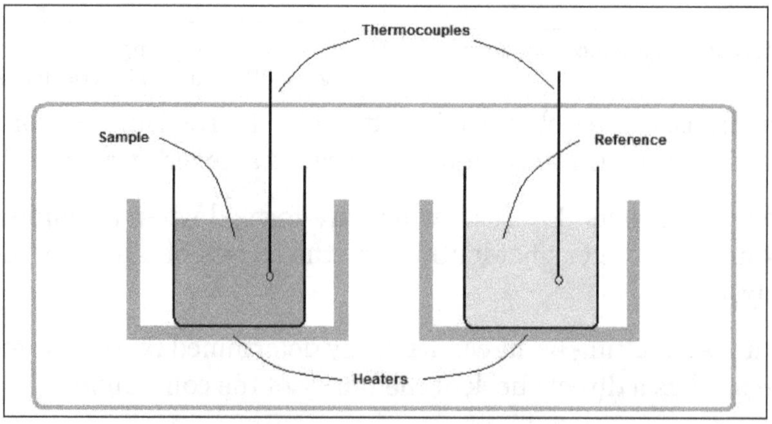

Differential scanning calorimeter schematic diagram.

For example, if the experimental goal is to determine the $\Delta_r H$ of a protein denaturation process, the reference cell could contain 100 mL H_2O, and the sample cell could contain 1 mg of the protein in addition to the same 100 mL H_2O. Therefore, the contribution of the solvent (H_2O) to the heat capacity of each cell would be equal, and the only difference would be the presence of the protein in the sample chamber.

Equations: The following equation relates the change in temperature to the change in enthalpy:

$$dH = \int_{T_i}^{T_f} nC_p dt$$

where dH is the rate of change in enthalpy, C_P is the molar heat capacity of the calorimeter, dT is the rate of change in temperature, n is the number of moles of material, and T_i and T_f are the initial and final temperatures, respectively. This equation can be integrated to yield,

$$\Delta H = nC_p \Delta T$$

where ΔH is the total change in enthalpy and ΔT is the change in temperature.

Differential Thermograms

The output yielded by differential scanning calorimetry is called a differential thermogram, which plots the required heat flow against temperature. Data analysis is highly dependent on the assumption that both the reference and sample cells are constantly and accurately maintained at equal temperatures. The graph below indicates the change in power (electrical heat) as the temperatures of the two cells are gradually increased. A change in specific heat results in a small change in power, and can be either positive or negative depending on the particular process. The advent of an endothermic reaction will cause an increase in power as temperature increases, since additional heat is required to drive the reaction and still maintain the reference temperature. When an exothermic reaction occurs, the opposite effect is observed; power decreases because heat is released by the reaction and less power is required to maintain equivalent temperatures in the chambers.

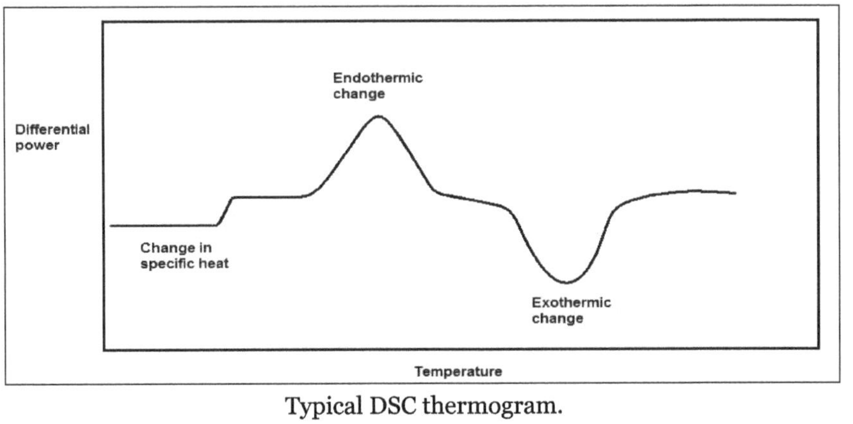

Typical DSC thermogram.

Differential scanning calorimetry can be used to study many different fields including biopolymer energetics where it is used to find the enthalpy of the protein denaturation process. A protein can be changed from its native state, in which it has a specific conformation due to non-covalent intramolecular interactions, to a denatured state where this characteristic structure is altered. Analysis of proteins through DSC can provide both the enthalpy of denaturation and information about the cooperativity of the denaturation process. A sharper peak in the thermogram indicates a higher level of cooperativity, meaning that when one structural association is disturbed, the likelihood of disruption at other points of association will be enhanced.

DSC is also used in conjunction with differential thermal analysis. Through the combination of these two techniques, thermal behavior of inorganic compounds can be studied while the melting, boiling and decomposition points of organic compounds and polymers are found.

DSC Applications

- Glass transition
- Melting points
- Crystallization times and temperatures
- Heats of melting and crystallization
- Percent crystallinity
- Oxidative stabilities
- Compositional analysis
- Heat capacity
- Heats of cure
- Completeness of cure
- Percent cure
- Purities
- Thermal stabilities
- Polymorphism
- Heat set temperatures
- Recyclates or regrinds

A schematic DSC curve demonstrating the appearance of several common features:

- Differential scanning calorimetry can be used to measure a number of characteristic properties of a sample. Using this technique it is possible to observe fusion and crystallization events as well as glass transition temperatures T_g. DSC can also be used to study oxidation, as well as other chemical reactions.
- Glass transitions may occur as the temperature of an amorphous solid is increased. These transitions appear as a step in the baseline of the recorded DSC signal. This is due to the sample undergoing a change in heat capacity; no formal phase change occurs.

- As the temperature increases, an amorphous solid will become less viscous. At some point the molecules may obtain enough freedom of motion to spontaneously arrange themselves into a crystalline form. This is known as the crystallization temperature (T_c). This transition from amorphous solid to crystalline solid is an exothermic process, and results in a peak in the DSC signal. As the temperature increases the sample eventually reaches its melting temperature (T_m). The melting process results in an endothermic peak in the DSC curve. The ability to determine transition temperatures and enthalpies makes DSC an invaluable tool in producing phase diagrams for various chemical systems.
- A schematic DSC curve of amount of energy input (y) required to maintain each temperature (x), scanned across a range of temperatures. Bottom: Normalized curves setting the initial heat capacity as the reference. Buffer-buffer baseline (dashed) and protein-buffer variance (solid).

The technique is widely used across a range of applications, both as a routine quality test and as a research tool. The equipment is easy to calibrate, using low melting indium for example, and is a rapid and reliable method of thermal analysis.

Liquid Crystals

DSC is used in the study of liquid crystals. As some forms of matter go from solid to liquid they go through a third state, which displays properties of both phases. This anisotropic liquid is known as a liquid crystalline or mesomorphous state. Using DSC, it is possible to observe the small energy changes that occur as matter transitions from a solid to a liquid crystal and from a liquid crystal to an isotropic liquid.

Oxidative Stability

Using differential scanning calorimetry to study the stability to oxidation of samples generally requires an airtight sample chamber. Usually, such tests are done isothermally (at constant temperature) by changing the atmosphere of the sample. First, the sample is brought to the desired test temperature under an inert atmosphere, usually nitrogen. Then, oxygen is added to the system. Any oxidation that occurs is observed as a deviation in the baseline. Such analysis can be used to determine the stability and optimum storage conditions for a material or compound.

Safety Screening

DSC makes a reasonable initial safety screening tool. In this mode the sample will be housed in a non-reactive crucible (often gold, or gold plated steel), and which will be able to withstand pressure (typically up to 100 bar). The presence of an exothermic event can then be used to assess the stability of a substance to heat. However, due to a combination of relatively poor sensitivity, slower than normal scan rates (typically 2-3°/min – due to much heavier crucible) and unknown activation energy, it is necessary to deduct about 75-100 °C from the initial start of the observed exotherm to suggest a maximum temperature for the material. A much more accurate data set can be obtained from an adiabatic calorimeter, but such a test may take 2–3 days from ambient at a rate of a 3 °C increment per half hour.

Drug Analysis

DSC is widely used in the pharmaceutical and polymer industries. For the polymer chemist, DSC is a handy tool for studying curing processes, which allows the fine tuning of polymer properties. The cross-linking of polymer molecules that occurs in the curing process is exothermic, resulting in a positive peak in the DSC curve that usually appears soon after the glass transition. In the pharmaceutical industry it is necessary to have well-characterized drug compounds in order to define processing parameters. For instance, if it is necessary to deliver a drug in the amorphous form, it is desirable to process the drug at temperatures below those at which crystallization can occur.

General Chemical Analysis

Freezing-point depression can be used as a purity analysis tool when analysed by Differential scanning calorimetry. This is possible because the temperature range over which a mixture of compounds melts is dependent on their relative amounts. Consequently, less pure compounds will exhibit a broadened melting peak that begins at lower temperature than a pure compound.

Food Science

In food science research, DSC is used in conjunction with other thermal analytical techniques to determine water dynamics. Changes in water distribution may be correlated with changes in texture. Similar to material science studies, the effects of curing on confectionery products can also be analyzed.

Polymers

DSC is used widely for examining polymers to check their composition. Melting points and glass transition temperatures for most polymers are available from standard compilations, and the method can show up possible polymer degradation by the lowering of the expected melting point, T_m, for example. T_m depends on the molecular weight of the polymer, so lower grades will have lower melting points than expected. The percentage crystallinity of a polymer can also be found using DSC. It can be found from the crystallisation peak from the DSC graph since the heat of fusion can be calculated from the area under an absorption peak. Impurities in polymers can be determined by examining thermograms for anomalous peaks, and plasticisers can be detected at their characteristic boiling points.

Metals

In the last few years this technology has been involved in metallic material study. The characterization of this kind of material with DSC is not easy yet because of the low quantity of literature about it. It is known that it is possible to use DSC to find solidus and liquidus temperature of a metal alloy.

Thermogravimetric Methods

Thermogravimetry is defined as "a technique whereby the weight of a substance, in an environment heated or cooled at a controlled rate, is recorded as a function of time or temperature". A plot of mass as a function of temperature (a thermogram) provides both qualitative and quantitative information. Thermogravimetry is of three types:

- Isothermal or static thermogravimetry: The sample weight is recorded as a function of time at constant temperature.
- Quasistatic thermogravimetry: Heating of sample is done to constant weight at each of a series of increasing temperature.
- Dynamic thermogravimetry: Heating of the sample is done in an environment whose temperature changes in a predetermined manner, usually at a linear rate.

Instrumentation

Instrument used for thermogravimetry consists of a precision balance (thermobalance) in which changes in temperature occurs at a linear rate. Representation of the results is done by a plot of weight change versus temperature or time. This plot is known as thermogravimetric curve (TG curve). Weight should be plotted on ordinate and temperature (T) or time (t) should be plotted on the abscissa.

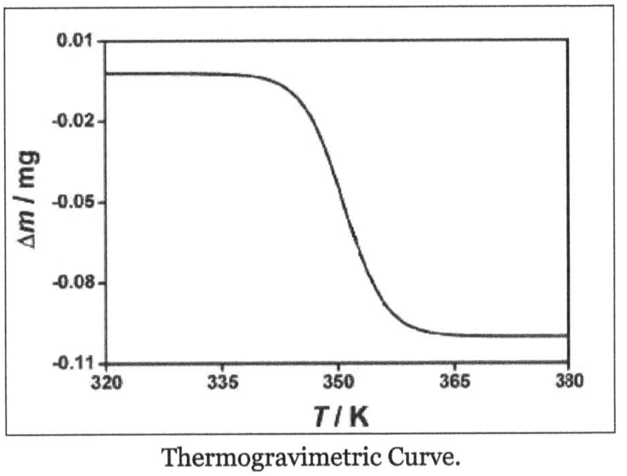

Thermogravimetric Curve.

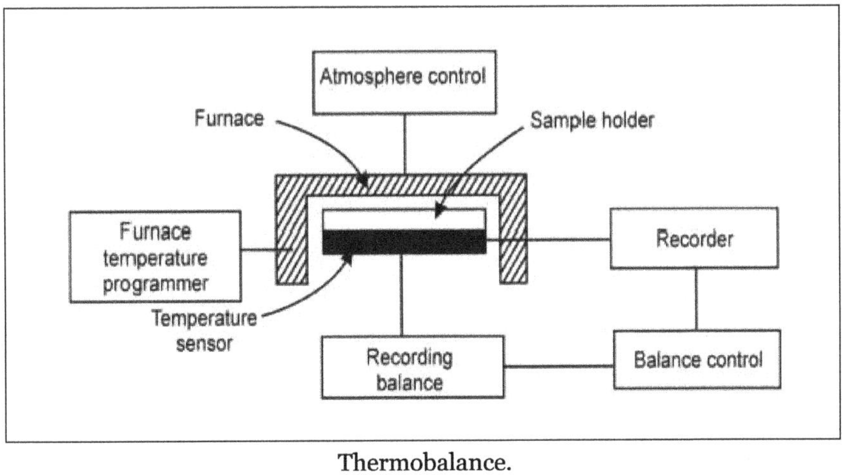

Thermobalance.

The principle of thermogravimetry is based on the simple fact that the sample is weighed continuously as it is being heated to elevated temperatures. In the market both manual as well as automatic recording balances are available but for practical reasons, the latter type is preferred. However, manual recording is still used for long term isothermal measurements (for example, with helix type thermobalance). Different components of a thermobalance are: recording balance, furnace, furnace temperature programmer and recorder. The specific details of each component depend on the particular application that is required of the instrument.

Basic Units of a Thermobalance

Balance

Balance is the most important component of a thermobalance. A good balance must possess following characteristics:

- Accuracy, reproducibility, capacity and sensitivity of a thermobalance should be similar to an analytical balance.
- Sufficient range of automatic weight adjustment should be present.
- Mechanical and electronic stability of a thermobalance should be high.
- Thermobalance should respond to weight changes rapidly.
- It should not get affected by vibration.
- It should be easy to operate.

Recording balances are mainly of two types: (i) deflection-type instrument and (ii) null-type instrument.

Deflection Balances

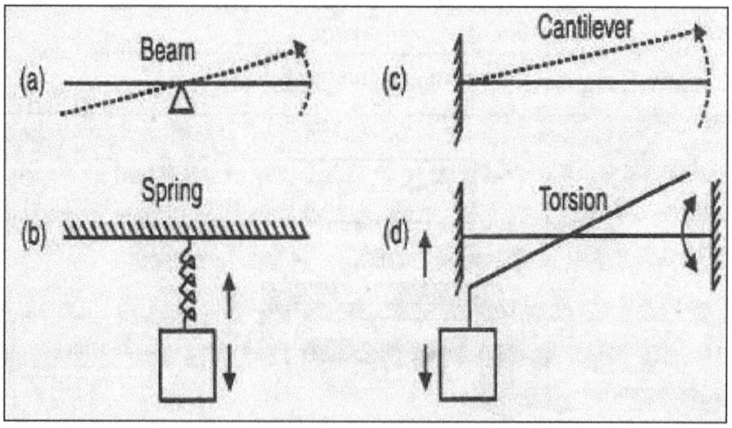

Deflection balances: (a) Beam type (b) Helical or spring type (c) Cantilever type (d) Torsion type.

There are of following types:

- Beam type: In these types of balances, conversion of beam deflection takes place about fulcrum into a suitable weight change which can be easily identified.
- Helical type: An elongation or contraction of the spring with weight change occurs in these types of balances. Transducers are used for suitable recording of changes in length of spring.

- Cantilevered beam: Here, one end of the beam is fixed while the other end where sample is placed undergoes deflections.
- Torsion wire: In these balances, a taut wire attached to the beam acts as a fulcrum. The taut wire is firmly fixed at ends such that deflection of beam is proportional to changes in weight and torsional characteristics of taut wire. Instead of a wire a ribbon made of metal can also be used.

Null–Point Balances

These types of balances are gaining much popularity in recent times in analysis of compounds. In null-point balances, a sensor is present which is useful in detection of the deviation of the beam from its null position. If deviation occurs then, a restoring force (electrical or mechanical weight loading) is applied to the beam for restoration of its null position from the horizontal or vertical norm. The force applied to restore the beam is proportional to change in weight and this restoring force is recorded directly.

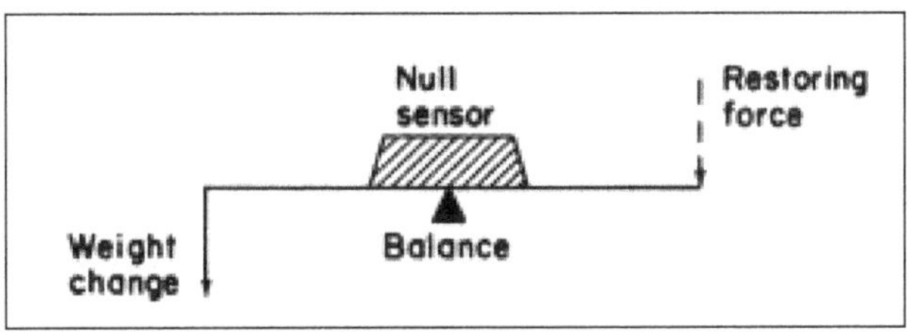

Null Type Balances.

Sample Holder for Placing Sample

Sample holder's geometry, size and material play a crucial role in determining the shape of a thermogravimetric curve. Sample holder's size and shape is governed by the nature of the sample, weight of the sample and the maximum temperature range to be used. Alumina, stainless steel, platinum, graphite, glass, quartz, etc. are used to make sample holders. There are following types of sample holders used to place samples in the furnace:

- Shallow pans: Shallow pans are usually employed in cases where it is necessary to eliminate diffusion as rate controlling step. As volatile material is produced throughout the sample mass, it should diffuse to the surface instantaneously to escape and be registered as a weight loss. In some cases, like polymers, side reactions may occur. In such cases, the sample is arranged in a thin layer so that volatile fragment is free to escape.
- Deep crucibles: Deep crucibles are used for samples where partial equilibrium or side reaction is wanted. These types of crucibles are employed to study calcinations at industrial scale. Deep crucibles are also employed in surface area measurements.
- Loosely covered crucibles: This type of crucible is usually employed in self-generated atmosphere studies.
- Retort cups: Retort cups resemble alchemist's retort. These types of cups prove to be very useful in determinations involving boiling points. The retort is helpful in providing the single plate of reflux necessary for a simple boiling point study.

Furnace used in Thermogravimetry

The furnace and control system should be designed to produce a linear heating rate over the whole working temperature range of furnace. The choice of furnace heating element and the type of furnace depend upon temperature being studied. For 1100 °C, the material of furnace is nichrome wire or ribbon. If a wire is being used, it should be wound coiled so as to accommodate differential thermal expansion of various components.

For temperature between 1100 and 1500 °C, one should use platinum or any alloy of platinum and rhodium. By using platinum – rhodium alloy having rhodium content of 40%, one can use the furnace up to 1750 °C. For temperature above 1750 °C, tungsten or molybdenum may be employed in a reducing atmosphere. The size of furnace is an important factor. It becomes easier to obtain a larger uniform hot zone in the high mass furnace. With the small mass furnace, it becomes difficult. The position of furnace relative to the balance is important. With some balances like quartz fibre spring balance, the furnace is below the weighing system but with beam balance, several choices are possible.

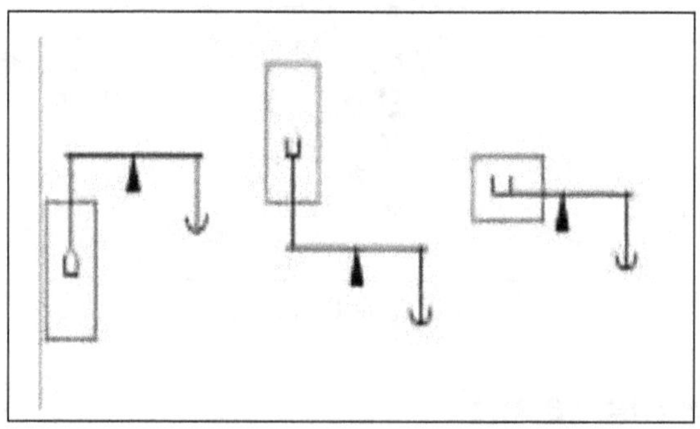

Position of furnace with respect to balance.

Measurement of Temperature

This measurement can be carried out in a number of ways. Thermocouple is the most common method used for temperature measurement. For measurement of temperature of about 1100 °C, chromal or alumel thermocouple constructed from alloys of platinum and rhodium are used. For very high temperatures, tungsten or rhodium thermocouples are generally used.

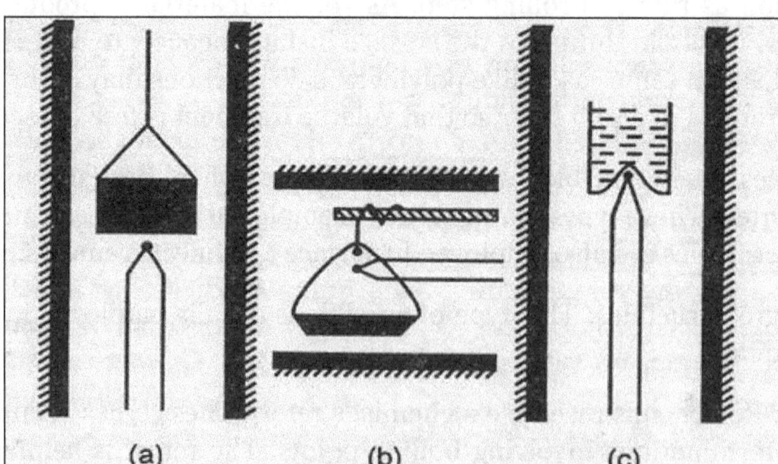

Position of a thermocouple in a thermobalance.

Recorder

The recording systems are mainly of two types:

- Time-base potentiometric strip chart recorder.
- X-Y recorders.

Thermobalance in Thermogravimetric Analysis

A good thermobalance should possess the following characteristics:

- The thermobalance should possess the capability to record continuously the changes in weight of a sample as a function of temperature or time.
- Furnace of thermobalance should have a wide range of temperature.
- Recording of temperature should be done to an accuracy of about ± 1%.
- The temperature which is recorded ideally is the temperature of the sample.
- Recording of loss in weight of the sample should be done to an accuracy of about ±1%.
- A linear heating rate should be used.
- The sensitivity of balance should be proportionate to the size of the samples being used.
- There should not be any chemical attack of volatile products on the apparatus.
- The crucibles should be located within hot zone.
- The balance has to be protected from the furnace due to its very high temperature.

Factors which Affect TG Curves

If study of a system is done by various thermobalances, then the shape of the thermogravimetric curve obtained with different thermobalances will vary from instrument to instrument. Thus, it is difficult to make meaningful comparisons between TG data obtained on different thermobalances in different laboratories. The difference in the shapes of TG curves may be attributed to different factors; leading to errors. The factors affecting the thermogravimetric curves can be divided into two main groups: Instrumental effects and Sample characteristics.

Thermogravimetry Applications

1. Automatic thermogravimetric analysis:

- Automatic gravimetric analysis for a single component system.
- Automatic gravimetric analysis for a binary mixture.

2. Evaluating precipitates obtained by gravimetry: By using thermogravimetric procedures one can easily determine the correct drying temperatures of various precipitates obtained in gravimetric methods. For example:

- Usually, for determination of lithium triple periodate is used. However, Duval in his study found out that it is not suitable for gravimetric analysis as the thermogravimetric curve of triple periodate showed that there occurs a constant decrease in weight from 40 °C to 947 °C and concluded that, there is no drying temperature for this precipitate.

- Duval rejected the salicylaldoxime method for determination of zinc as he was unable to find any constant weight plateaus because he employed a fast heating rate. The workers confirmed that 135 °C is the minimum drying temperature. Later Kettaach proved that a temperature of about 150 °C is sufficient for drying of the complex, regardless of the moisture content.

3. Evaluating suitable standards: Duval carried out a number of experiments on thermogravimetry to find suitable standards in analytical chemistry. Duval combined his TG studies with IR absorption spectrophotometric studies order to arrive these conclusions. He confirmed that the following substances cannot be used for preparing standard solution: Magnesium ammonium chloride, ammonium bicarbonate and ammonium fluoride. He observed that following compounds are most suitable for preparing standard solutions: Lithium sulphate monohydrate, Sodium dichromate dehydrate, Sodium cobaltinitrate, Hydrazinium chloride, Hydrazinium sulphate, Ascorbic acid and methylglucamine. Duval advised that not to heat the compounds above the temperature given in the following table:

Substances	Temperature °C
EDTA	109
Urea	165
NaF	850
$Ni(NO_3)_2.6\ H_2O$	50
$KHCO_3$	125
$NiSO_4.7H_2O$	100

4. Testing purity of samples: As an analytical tool, the thermobalance can be used in determining the purity of various substances.

5. Determining Curie point: If a ferromagnetic material is kept on a thermobalance with the pole of the magnet above the sample, the TG balance will show the weight which will be less than the actual weight. Ferromagnetic materials lose their magnetism on heating at exactly reproducible temperature on curie points. A range of metals or alloy with curie points between 150 °C and 1000 °C is available. In the TG curve, the temperature at which the thermogram shows an increase in weight is the curie temperature of this substance.

6. Organic substances: Many organic substances have been successfully studied by DTA and TG methods.

7. Oxide mixture: Thermogravimery has been successfully employed to study the reaction between various oxides. For example, the system $CuO\text{-}Cu_2O\text{-}Cr_2O_3$ was studied by this technique.

8. Building materials: One of the interesting examples is cement where study of a ternary compound in the system $CaO\text{-}\ Al_2O_3\text{-}SO_3$ is done.

Derivative Thermogravimetry (DTG)

In thermogravimetric analysis, mass of a sample (m), is continuously recorded as a function of temperature (T) or time (t), i.e.,

$$m = f(T \text{ or } t)$$

The quantitative measurement of the change in mass is done by determining the distance, on the curve mass axis, between the two points of interest or between two horizontal mass levels.

In derivative thermogravimetry, derivative of the mass change with respect to time is recorded as a function of temperature or time,

$$\frac{dm}{dt} = f(T \text{ or } t)$$

The curve which an analyst obtains is the first derivative of the mass change curve. Instead of the stepwise curve a series of peaks are obtained. And the areas under the peaks are proportional to the total mass- change of the sample. A derivative and a mass loss curve are shown in the figure.

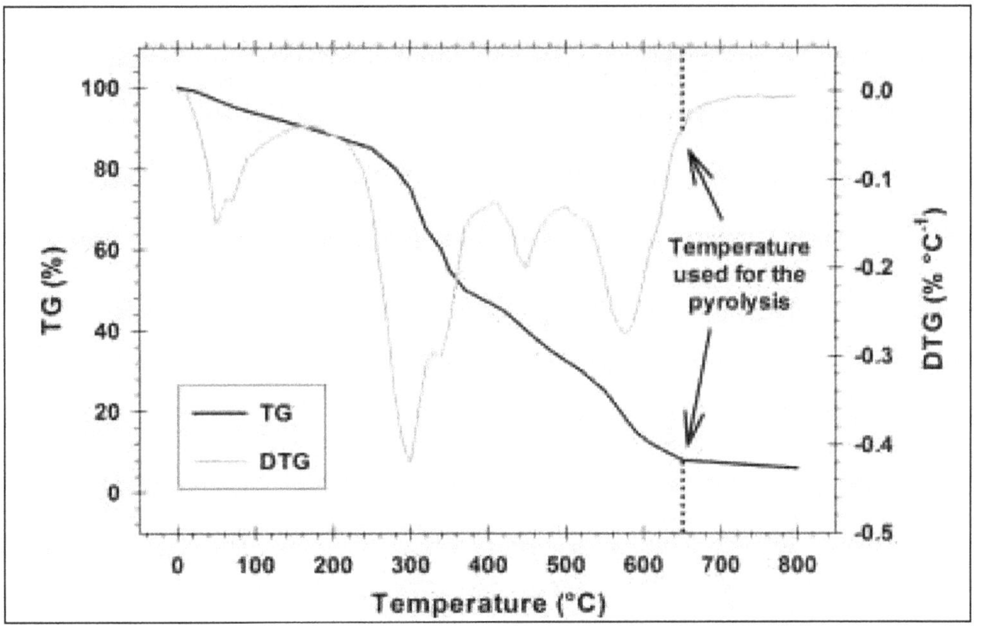

A TG and DTG mass loss curve.

The derivative curve may be obtained either from thermogravimetric curve by manual differentiation methods or by electronic differentiation of the TG signal. Accessory equipment is available for most thermobalances so that the derivative curve can be recorded, without any difficulty, along with the thermogravimetric curve. It should be noted that the DTG and TG curves are obtained under similar experimental conditions. Figure shows a thermogram of nylon and its DTG curve. There is a minimum in the DTG curve at the point of inflection of the thermogravimetric curve.

This makes identification simpler. Consequently, DTG is a valuable method of data presentation for qualitative analysis.

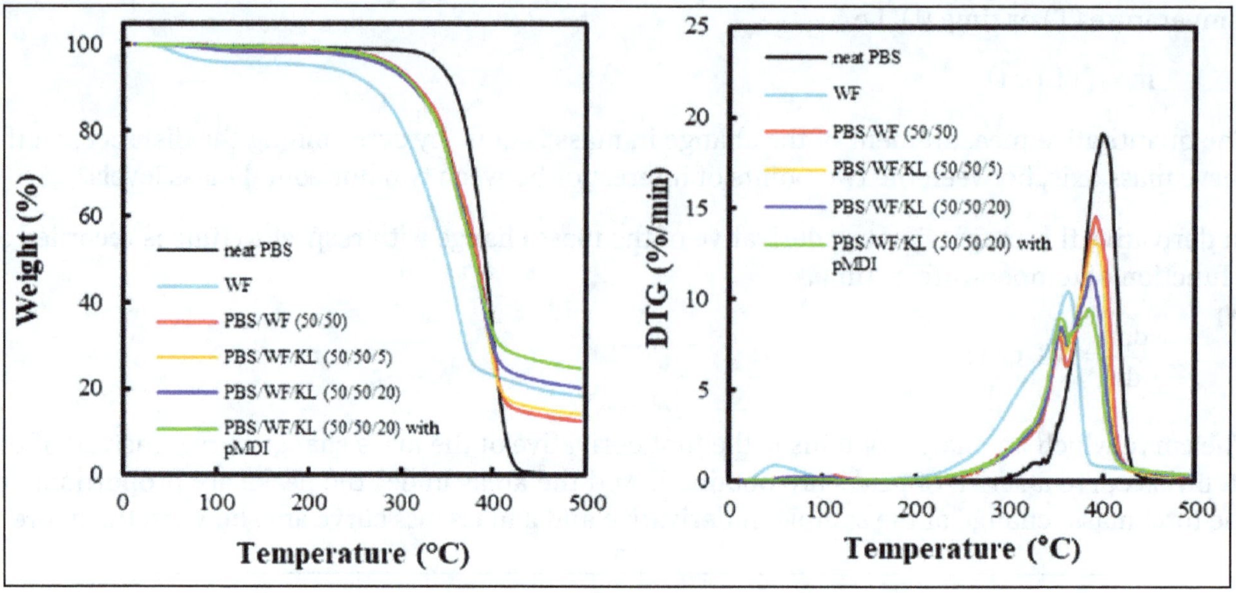

Thermogravimetric analysis (TGA) and derivative thermogravimetry (DTG) curves for the PBS/WF and PBS/WF/KL composites with and without pMDI.

References

- Differential-Scanning-Calorimetry, Physical-and-Theoretical-Chemistry-Textbook-Maps, thermodynamics, calorimetry: libretexts.org, Retrieved 10, August 2020
- Thermal-analysis-applications: instrument-specialists.com, Retrieved 22, February 2020

5

An Integrated Study of Electroanalytical Methods

The methods which are used to study an analyte by calculating the potential and current of an electrochemical cell holding the analyte are known as electroanalytical methods. Science and technology have undergone rapid developments in the past decade which has resulted in the discovery of significant electroanalytical methods; which have been extensively detailed in this chapter.

Chemical reactions, in which one or more electrons are transferred into one species from another, provide the basis for electroanalytical chemistry. Here we deal with electrical quantities, such as volts, ohms, amperes, and their relationship to chemical and physical properties of the species in a given system. Electrical quantities can be measured precisely and instantaneously with readily available instrumentation that may be automated or used to control the reaction being investigated. Here we shall discuss about kinds of electrodes, the factors that determine their potentials, and some of the factors that determine the current flowing through the cell. Due to relative inertness of water and high dielectric constant it is the most satisfactory solvent. Although many organic solvents are also suitable. By far the most common use of electrochemical cells in chemical analysis is the determination of the concentration of one of the species in a solution, for example, the determination of hydrogen ion with the pH meter.

There are many ways in which an electrochemical cell can be constructed and operated giving rise to a group of techniques for special purposes. Some of the important electroanaylytical techniques are:

- Potentiometry: The potential of an electrode is used to measure the concentration of a given species in a non-reacting system.
- Potentiometric titration: The concentration of a given species that changes during a titration is measured by potentiometry.
- Electrolysis: The amount of a substance present is determined by reducing or oxidizing it at an electrode, measuring either changes in weight of the electrode (electrodeposition) or the number of coulombs required to complete the reaction (coulometry).
- Coulometry: Electrolysis is performed at constant current or with one of the electrodes at a fixed potential and the number of coulombs measured.
- Voltammetry: Electrolysis is performed in an unstirred solution and the current is measured as a with respect to applied potential, or vice versa.
- Polarography: A voltammetric method in which one of the electrodes is an easily polarized microelectrode (usually a dropping mercury electrode) and the current is measured with respect to the applied potential and time.
- Chronopotentiometry: Electrolysis is performed in an unstirred solution at constant current and the potential of an easily polarized electrode is measured as a function of time.

- Conductometry: The electrical resistance of a solution is used as a measure of the total ionic concentration in the solution.

Although some of these methods can no longer be considered "modern," applications in the areas of organic chemistry, biochemistry, and pollution monitoring continue to increase.

Electrochemical Cells

The reaction between zinc metal and ferric ion is a typical redox reaction which can be carried out by dipping a piece of Zn metal in the solution of ferric ion, or by setting up an electrochemical cell as shown figure in which this reaction takes place. One half-cell contains a solution of a zinc salt and a piece of metallic zinc; the other contains, solution of Fe^{2+} and Fe^{3+} ion and a piece of platinum metal that serves as an inert electrode to supply or withdraw electrons from the solution.

In the zinc half-cell, zinc metal may dissolve as zinc ion, leaving the electrons behind to give the electrode a negative charge. Alternatively, zinc ion may deposit as zinc metal on the electrode, causing a deficiency of electrons on the electrode. For a given concentration of zinc ion, an equilibrium situation is established with a resultant charge on the electrode. Likewise in the ferrous-ferric half-cell, ferrous and ferric ions can exchange electrons with the platinum electrode establishing equilibrium and a resultant charge on the platinum electrode.

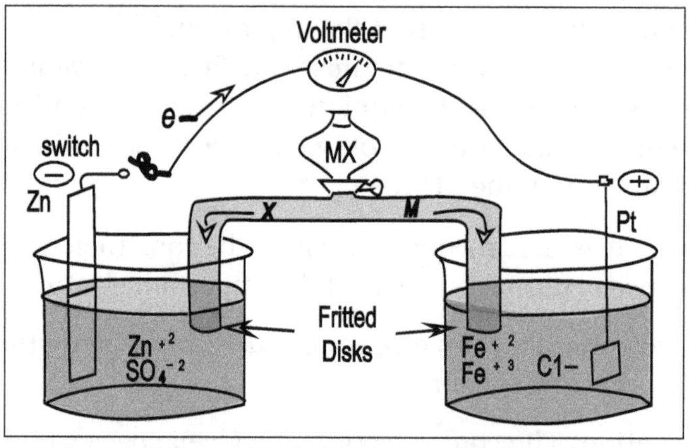

Schematic representation of Cell 1.

If the two electrodes are connected with a wire, electrons must flow from zinc to platinum and the reaction just described presumably will occur. However, we cannot increase the concentration of zinc ion or change the initial ratio of ferrous to ferric ion without disturbing the electro neutrality of the solutions. Since we cannot accumulate electrical charge in either solution, the reaction cannot continue unless we provide a mechanism for transferring ions from one half-cell to the other. The salt bridge serves this function by allowing a small ion current to pass from one half -cell to the other in either direction with only negligible mixing of the two electrolytes.

When this cell operates spontaneously: (a) zinc metal is oxidized and dissolves, (b) electrons are transferred from the zinc electrode to the platinum electrode through the wire, (c) ferric ion is reduced by the electrons on the platinum electrode, (d) cations migrate from the zinc half-cell to the ferrousferric half-cell through the salt bridge, and (e) anions migrate in the reverse direction. By means of appropriate switches and resistances in the external circuit we can start or stop the cell reaction or control its rate. By means of appropriate meters we can measure its rate and the driving force.

The operation of a cell clearly requires a chemical reaction to take place. Thus the concentrations of the various species will necessarily change with time. For the most part we will be concerned only with cells in which no current is flowing. It is important to note that the charges on the electrodes are the result of chemical reactions; however, the amount of chemical reaction required to establish the charge is so slight that it cannot be measured and is safely neglected when compared to the total material in the cell.

For example, a free charge of 10^{-17} moles of electrons will cause a potential of 1 volt at a distance of 1 cm. In these no operating cells, we shall assume that each half-cell reaction is in equilibrium with its own electrode. A variance in potential between two electrodes implies that the two half -cell reactions are not in equilibrium with each other. In fact, the potential difference measures the force tending to drive the system toward equilibrium.

Some Typical Electrodes

A, a half-cell is designated by listing the species taking part in the half-reaction, separating the electrode from the electrolyte by a vertical bar. A few examples will illustrate some of the conventions.

- Metal-metal ion, in which the metal of the electrode takes part III the half-reaction:

  Zn/Zn^{+2}

- Metal-complex ion, in which an excess of the complexing agent is added to regulate the particular form of the complex:

  $Cu \mid Cu\,(NH_3)_4^{+2}, NH_3$

- Metal-saturated solution of one of its salts, in which the presence of the solid salt is inconsequential except to insure that the solution is saturated:

  $Ag \mid AgCl, Cl^-$

- Gas-ion, in which the solution is saturated at given pressure with the gas as it is bubbled over the surface of inert metal electrode:

  $Pt, H_2 \mid H^+$

- Ion-ion, in which both the reduced and oxidized forms are soluble in the electrolyte, and an inert metal electrode (usually platinum) is used to make electrical contact between the electrolyte and the external circuit:

  $Pt \mid Fe^{+3}, Fe^{+2}$

- Amalgam-ion, in which a reactive metal is amalgamated so that direct chemical reaction with the solvent is minimized:

  $Na(Hg) \mid Na^+$

The order in which substances in the electrolyte are listed is irrelevant. When necessary to specify concentrations, this may be done within parentheses:

$Pt \mid Cr^{+3}(0.5\ M), Cr_2O_7^{-2}(0.1\ M), H^+(1.5\ M)$

Finally, if two half-cells are combined to make a whole cell, the electrodes are listed first and last; for example, the cell in figure would be given as:

Cell 1: $Zn \mid Zn^{+2} \| Fe^{+3}, Fe^{+2} \mid pt$

The double bar signifies the salt bridge between the two half-cells. Unless otherwise indicated, the solvent is assumed to be water and it is not listed as one of the cell ingredients even though it may take part in the reaction. In so far as practical, we shall write the predominant form of a species in the solution; for example, Fe^{+3}, $Fe(OH)^{+2}$, $FeCl_2^{+2}$, HOAc, OAc^-, etc.

The charge on an electrode results from an excess or deficiency of electrons on the metal. A large negative charge indicates the presence of a strong reducing agent (good electron donor). The potential of an electrode is defined in an electrostatic sense. We shall be concerned only with potential differences between two points; namely, the two electrodes of a cell. We should note that the point with the higher electric potential also has the higher positive charge. The potential difference thus compares the charge existing at two points, a relative measurement. We often use the term, "potential of an electrode," when we mean potential difference relative to an arbitrary standard reference electrode. Likewise, a "cell potential" is really the difference in potential between the two electrodes, and is more properly called the cell voltage or electromotive force. We shall designate this term as E_{cell} rather than by ΔE.

Sign Conventions

The charge on an electrode, the electrode potential, the difference between two electrode potentials (cell voltage) and the direction of electron flow in a wire connecting the electrodes are all physical facts, independent of any arbitrary sign conventions. Here, we shall use following rules:

- The potential of and charge on an electrode and the driving force of the corresponding half-reaction have the same numerical value and sign. Thus, the electrode potential, designated by the symbol E, is also a measure of:
    - The charge on the electrode.
    - The tendency of the system to accept electrons, that is, its power to oxidize some other species.
- E for an electrode or a half-reaction is compared to and measured relative to E for the standard hydrogen electrode (SHE) to be described shortly. Thus the charge on the electrode and the oxidizing power of a system may be either positive or negative relative to the SHE.
- In every cell, one electrode is arbitrarily designated as the left electrode, E_L, and the other as the right electrode, E_R.
- The cell voltage, E_{cell} is the algebraic difference between the two electrodes, always taken in the direction: $E_R - E_L = E_{cell}$.
- All the half-reactions are written as reductions (electrons on the left).
- The cell reaction is obtained by subtracting the left half -cell reaction from the right. It

should be evident that the cell reaction cannot produce or consume electrons, it can only transfer them. If the number of electrons, n, is not the same in the two half-reactions, one or both half-reactions must be multiplied by a suitable factor to cancel electrons. Note that multiplying or dividing a half-reaction by a numerical factor does not change the electrode potential, a property which is independent of how the half-reaction is written.

- A positive cell voltage shows that:
  - The cell reaction proceeds spontaneously in the direction written.
  - The right electrode has a more positive (or less negative) potential than the left.
  - Reduction occurs in the right half-cell and oxidation in the left when the cell operates spontaneously.
  - The flow electrons are from left to right in the external circuit.
- A negative cell voltage corresponds to statements opposite to those just given.

Standard Hydrogen Electrode

No one has yet measured a single electrode potential. Nor is it likely that anyone will ever do so, because a second electrode is always necessary in order to complete the electrical circuit. The voltmeter (or other voltage measuring device) can measure only differences in electrode potentials. But we can assign an arbitrary value to a selected electrode. If this so-called "primary standard" electrode is connected with some other electrode, we can then measure the potential of the second electrode with respect to the standard. Chemists have generally agreed to the arbitrary assignment of zero volts to the standard hydrogen electrode (SHE). Actually, this is a hypothetical electrode which is illustrated in Figure. Any other electrode, once it has been calibrated against the SHE, will serve as a secondary standard.

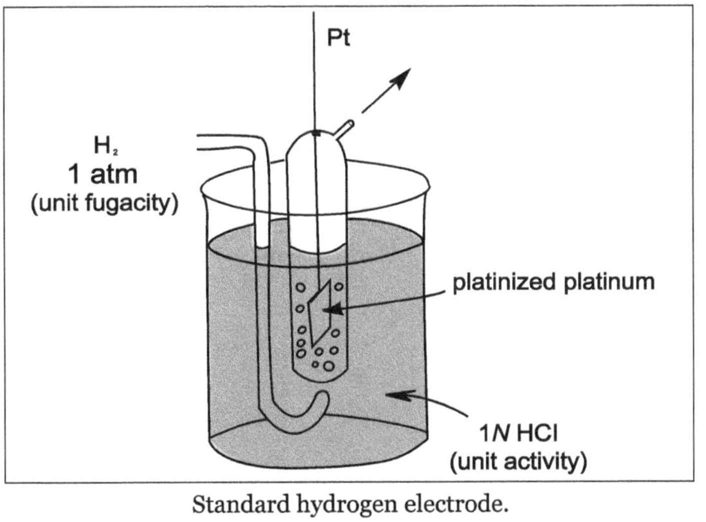

Standard hydrogen electrode.

Effect of Concentration

Let us consider the ferrous-ferric electrode in more detail. In Figure, the ferric ions and ferrous ions are in continuous motion and collide frequently with each other and with the electrode surface. For the ferric ions which reach the electrode surface, there is a finite probability that they will "pick up" an electron and become ferrous ions. The degree of the probability depends on the electron affinity of the ferric ion and the activity of electrons at the electrode surface. Similar considerations apply to the ferrous ions. Both processes occur simultaneously and for the most part cancel each other out. However, the equilibrium condition can be described by the ratio of the activities of the two ionic species and the activity of electrons at the surface of the electrodes. The latter is more easily described as the charge or potential of the electrode. The mathematical expression that represents the equilibrium situation is called the Nernst equation:

$$E = E^{o}_{Fe^{+3},Fe^{+2}} - \frac{RT}{nF}\text{In}\frac{(Fe^{+2})}{(Fe^{+3})}$$

In which $E^{o}_{Fe}{}^{+3}.\ Fe^{+2}$ is a constant at a given temperature known as the standard potential, R the gas constant, T the absolute temperature, n the number of electrons transferred in the half-reaction and F the Faraday constant. The Nernst equation can be derived from thermodynamic considerations or it can be obtained empirically as the best fit to the experimental data.

If the activity of Fe^{+2} is increased, there is a greater probability for their oxidation at the electrode surface, a higher density of electrons accumulates, and the electrode becomes more negative. The opposite is true for an increase in the ferric ion activity. However, it is the ratio of activities that is important.

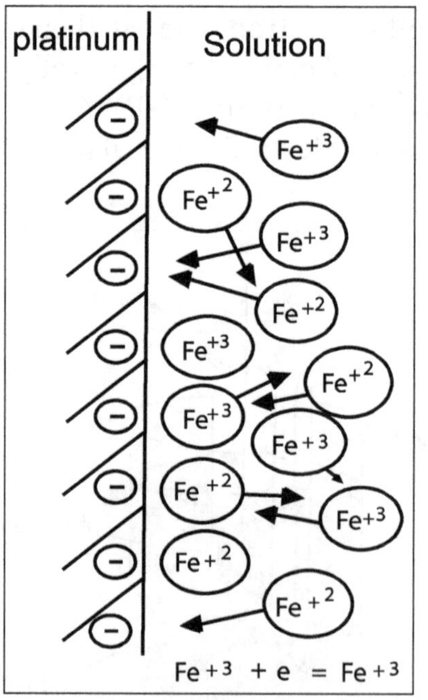

The dynamic equilibrium at the ferrous-ferric electrode.

Standard Potentials

The constant term, E°, is the observed potential of the electrode when the activities of the two species, ferric and ferrous ion, are both unity; in which case the log term in Equation

$E = E^{o}_{Fe^{+3},Fe^{+2}} - \frac{RT}{nF} \text{In} \frac{(Fe^{+2})}{(Fe^{+3})}$ is zero. E° expresses the inherent tendency of the system to accept electrons from the electrode, and its value depends on the nature of the substances in the half-reaction. More accurately stated, the standard potential is the measured potential of an electrode (relative to the SHE) when all participating species are in their "standard states."

Customarily, standard states for electrochemistry are defined as follows:

- For all dissolved substances, activity equal to one molar.
- For all gases, partial pressure equal to one atmosphere (more precisely, unit fugacity).
- For all pure liquids and pure solids, the most stable form at 25 °C, which by definition has unit activity.

In principle, standard potentials are measured by means of cells such as cell 2:

$$Pt, H_2 | H^+ | | Cu^{+2} | Cu$$

With all species in their standard states; or by comparing the electrode to any secondary reference electrode which has been previously compared to the SHE.

In practice, this procedure is not as simple as it may seem. Some of the complicating factors to be considered below: (a) the difficulty of obtaining truly reversible electrode reactions, (b) the uncertainty of the junction potential between the two half-cells, and (c) non-ideal behavior of solutes. Some E° values cannot be measured at all and are computed indirectly from other kinds of related thermodynamic data.

Use of the Nernst Equation

It is convenient to combine the constant factors in the log term of equation $E = E^{o}_{Fe^{+3},Fe^{+2}} - \frac{RT}{nF} \text{In} \frac{(Fe^{+2})}{(Fe^{+3})}$ and convert natural logarithms (base e) to common logarithms (base 10). The simplified form of the Nernst equation is:

$$E = E^{o} - 0.059/n \log (\text{products})/(\text{reactants})$$

Where 0.059 is valid for 25 °C, 0.060 at 30 °C, etc. The log term is set up much like the mass action expression for the reaction, but the numerical value is not equal to the equilibrium constant. Nernst equations for some of the electrodes used in cells 1 to 7 will serve as examples.

Activity Coefficients

Strictly speaking, the Nernst equation requires the use of activities rather than concentrations. The activity or effective concentration differs from the actual concentration because of interionic attractions and repulsions, as well as the finite space occupied by the various species and their solvation shells. To some extent, thermal motions modify the electrical effects. The activity coefficient expresses the ratio of activity to concentration,

$$f_i = a_i / C_i \quad \text{or} \quad a_i = f_i C_i$$

Where the subscript "i" stands for any given species. Although some may think f stands for "fudge factor," actually we know a great deal about this factor that measures the departure of real systems from ideal behavior. There is, however, no way to be sure that we have included every possibility. Consider the problem of the activity coefficients in the following cell:

Cell 8: pt, $H_2 \mid H^+, Cl^-, AgCI \mid Ag$

This cell requires only a single electrolyte with both electrodes immersed in the same solution of HCl - there is no salt bridge. The voltage for this cell is obtained from the two half-cell potentials:

$$\text{Left: } E_L = E_H{}^{\circ} - 0.059/2 \log P_{H2} / \left(H^+\right)^2$$

$$\text{Right: } E_R = E^{o}{}_{AgCl,Ag} - 0.059 \log \left(Cl^-\right)$$

$$E_{cell} = E^{o}_{AgCl,Ag} + \frac{0.059}{2} \log P_{H_2} - 0.059 \log\left(H^+\right)\left(Cl^-\right)$$

Here we have the product of two activities. It is impossible to construct a cell in which there is only one ion in the cell reaction. While we can measure the concentration of each ion (for example, by titration or precipitation) we can measure only the product of their activities and thus only the product of their activity coefficients. If both ions have the same numerical charge, it is customary to assign the same value to each activity coefficient. This clearly must be the square root of the product, or the geometric mean. The mean ionic activity coefficient is thus defined:

$$f_{\pm} = \sqrt{f_+ \times f_-}$$

Although the mean ionic activity coefficient is the only one we can measure, the Debye-Huckel limiting law provides a means to estimate the value of an individual ionic activity coefficient.

$$-\log f_i = 0.512\ Z_i^2 \sqrt{\mu}$$

Where Z_i is the charge on the ion and p, is the ionic strength of the solution. Equation above as given applies reasonably well to dilute aqueous solutions with concentrations no greater than about 0.05 M. Equation above has been modified by many workers in an effort to approach the behavior of real systems at higher concentrations. One such equation takes into account the effect of the finite size of the ions.

$$-\log f_{\pm} = 0.512 Z_i^2 \left[\frac{\sqrt{\mu}}{1 + Ba\sqrt{\mu}}\right]$$

Where B is a function of the dielectric constant and the temperature (B = 0.328 for water at 25 °C) and is an ion size parameter. Equation above extends the applicability of the Debye-Huckel limiting law up to concentrations of 0.1M.

The values depend on the charge and size of the ion, and on the ionic strength, μ, of the solution which is equal to one-half of the sum of the concentration of each ion multiplied by the square of its charge,

$$\mu = \frac{1}{2}\Sigma C_i Z_i^2$$

The definition of activity coefficients, Equations $f_i = a_i / C_i$ or $a_i = f_i C_i$ and $f_\pm = \sqrt{f_+ \times f_-}$, is used in rewriting equation $E_{cell} = E^o_{AgCl,Ag} + \frac{0.059}{2}\log\ P_{H_2} - 0.059\log\left(\mathrm{H}^+\right)\left(Cl^-\right)$:

$$E_{cell} = E^o_{AgCl,Ag} + \frac{0.059}{2}\log P_{H_2} - 0.059\log C^2_{HCl} - 0.059\log f^2_\pm$$

All measurable terms are gathered on the left:

$$E_{cell} = E^o_{AgCl,Ag} + \frac{0.059}{2}\log P_{H_2} - 0.059\log C^2_{HCl} - 0.059\log f^2_\pm$$

It is found from experiment that a plot of the left side of equation above versus √μ becomes nearly a straight line at low ionic strength, At zero concentration, or zero ionic strength, log f± also becomes zero, and the intercept gives the value of $E^o_{AgCl,Ag}$. This is the fundamental method of measuring E° values. Now that E° has been determined we can use equation above or Figure to determine $f_\pm$ for HCl at any ionic strength.

Liquid Junction Potential

Cell 8 is a special type of cell with only a single electrolyte and therefore no liquid junction (salt bridge). Most cells require two electrolytes with a junction between them. This introduces another uncertainty into our measurements. There will be more chances of diffusion of species from one solution when the solutions of different compositions are in contact, there is a tendency for to the other in order to equalize the concentrations. In general, the diffusion rates of the various species will not be the same and this will cause a slight separation of charges near the boundary between the two solutions. The electrical potential generated by this separation is called the junction potential, E_j it will, of course, be included in the measured cell potential.

In a few simple cases, the magnitude of this junction potential can be calculated from known diffusion coefficients. In general, the problem is too complex to handle. The junction potential can be reduced (but not altogether eliminated) by the use of a saturated solution of KCl in the salt bridge.

Reversibility

In the discussion of the Nernst equation, we have tacitly assumed that the electrode system behaves reversibly. By this we mean that the components within the half-cell system are always at equilibrium with the electrode. In other words, an infinitesimal change in the charge on the electrode results in an immediate change in concentrations to maintain the equilibrium; or, conversely, that an infinitesimal change in the concentrations results in an immediate change in charge on the electrode. This represents an ideal system; real systems approach this behavior to varying degrees.

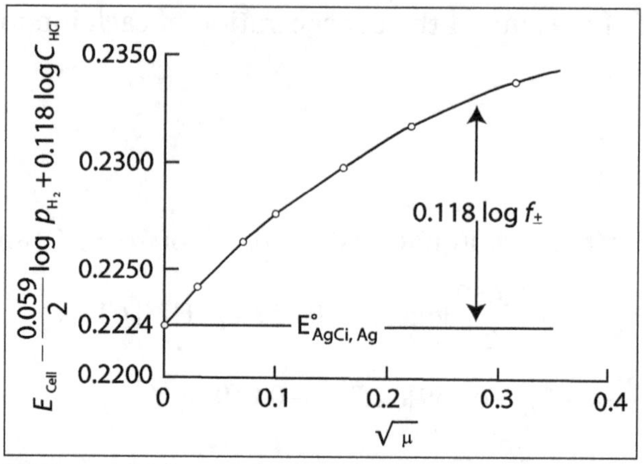

Plot of Equation IS-IO to yield the standard potential of the silver-silver chloride electrode and the mean ionic activity coefficient of HCl solutions.

A few of the soft metals (Ag, Hg, Pb, Zn, Cu, etc.) with their respective ions meet this criterion very well. Some others behave reversibly if the metal is amalgamated with mercury (Na, Cd, Bi, Tl, etc.). On the other hand, the mechanisms of most electrode reactions are very complicated indeed. The transfer of electrons between the electrode and electroactive species in solution can take place rapidly and reversibly only when the reduced and oxidized forms have essentially identical structures. Any rearrangement in the solvation shell or a transfer of atoms from one species to another will require promotion of the species to a higher energy state (activated complex). Under these circumstances the electrode reaction may take place very slowly and irreversibly or not at all, and the Nernst equation will be a poor approximation of the observed behavior. It is not surprising that many inorganic and most organic electrode reactions are irreversible. Many of these do, however, give reproducible behavior and are useful on an empirical basis.

Formal Potentials

Many redox reactions involve species other than those between which electrons are transferred. For example, many oxidants contain oxygen, requiring that hydrogen or hydroxyl ions and water appear in the reaction. In other cases, complexing agents may be consumed or released. Consider, for example, the reaction between copper (I) and permanganate in hydrochloric acid:

$$10Cl^- + 5CuCl_2^- + MnO_4^- 8H^+ \rightleftharpoons 5CuCl_4^{-2} + Mn^{+2} + 4H_2O$$

We can imagine this as a cell reaction and write each half-cell potential separately:

$$E_{Mn} = E^o_{MnO_4^-, Mn^{+2}} - \frac{0.059}{5}\log\frac{\left[Mn^{+2}\right]}{\left[MnO_4^-\right]\left[H^+\right]^n}$$

$$E_{Cu} = E^o_{CuCl_4^{-2}, CuCl_2^-} - 0.059\log\frac{\left[CuCl_2^-\right]\left[Cl^-\right]^2}{\left[CuCl_4^{-2}\right]}$$

When we use this reaction, we employ an excess of hydrochloric acid and its concentration stays relatively constant as the reaction proceeds. Under these conditions, [H$^+$] in equation

$E_{Mn} = E^{o}_{MnO_4^-,Mn^{+2}} - \frac{0.059}{5}\log\frac{[Mn^{+2}]}{[MnO_4^-][H^+]^n}$ is a constant and the log term can be separated:

$$E_{Mn} = \underbrace{E^{o}_{MnO_4^-,Mn^{+2}} + \frac{0.059}{5}\log[H^+]^8}\; - \frac{0.059}{5}\frac{[Mn^{+2}]}{[MnO_4^-]}$$

$$E_{Mn} = E^{'}_{MnO_4^-,Mn^{+2}} - \frac{0.059}{5}\log\frac{[Mn^{+2}]}{[MnO_4^-]}$$

Above equations define the formal potential, E', for this couple. E' is a constant which has a different value for each concentration of hydrogen ion. E' can be described as an effective or practical value of EO, and in this case it is seen that E' becomes more positive by (8/5) x 0.059 V as $[H^+]$ is increased tenfold. Thus changing the pH is a simple and powerful means for altering the oxidizing or reducing power of a reagent. This is a common situation for many inorganic and most organic reactions.

The same argument can be applied to define a formal potential for the copper half-cell which includes the effect of the concentration of Cl^-:

$$E_{cu} = \underbrace{E^{o}_{CuCl_4^{-2},CuCl_2^-} - 0.059\log[CI^-]^2}\; - 0.059\log\frac{[CuCI_2^-]}{[CuCI_4^{-2}]}$$

$$E_{Cu} = E^{'}_{CuCI_4^{-2},CuCI_2^-} - 0.059\log\frac{[CuCl_2^-]}{[CuCl_4^{-2}]}$$

Formal potentials include the effects of ionic strength, complexation, hydrolysis, the pH effect and any other effect which can be attributed to the composition of the electrolyte other than the ratio of reduced to oxidized forms.

Types of Electrodes

Electrodes are classified in two broad groups: indicator electrode and reference electrode. A reference electrode should easy to construct, and need to maintain a constant, reproducible potential even if small currents are passed through it. An indicator electrode must respond to changes in activity of the species to be measured. Some electrodes are especially useful for one or both of these purposes.

The Hydrogen Electrode

Although this electrode is the standard of reference, it is inconvenient to construct and difficult to maintain. A source of highly purified hydrogen is required, and its partial pressure within the cell must be regulated. The platinum surface is coated with a thin layer of platinum black to increase its surface area and catalyze the electrode reaction. Traces of impurities in either the gas or solution may poison the electrode. The electrode responds to most other redox couples and therefore strong oxidants or reductants must not be present. Its use is limited mostly to calibrating other electrodes and to measuring hydrogen ion activities for solutions in which no other electrode will operate satisfactorily.

The Quinhydrone Electrode

The composition and function of this electrode have just been discussed, and it is only necessary to add that a solution containing equal concentrations of quinone and hydro quinone is easily prepared from solid quinhydrone, a molecular addition compound of quinone and hydroquinone:

$$2QH \rightleftharpoons Q + H_2Q$$

Previous equations then becomes,

$$E = 0.6994 - 0.059\ pH$$

Hydroquinone may be oxidized in the presence of strong oxidants and it will dissociate as a weak acid $(pK_a \sim 10)$ in appreciable amounts above pH 8.5. Either of these processes will disturb the ratio of $[H_2Q]/[Q]$ and give erroneous results for pH.

Smooth Platinum Electrode

A piece of smooth, bright platinum wire or plate responds to most redox couples when all species are present in the solution (including dissolved gases). In moderately strong acid solutions, we would expect the platinum to respond to pH (as in the SHE). However, with smooth platinum, the reduction of hydrogen ion is somewhat irreversible and the electrode is normally insensitive to pH. Very strong reductants, for example, Cr^{+2}, may be oxidized by hydrogen ion in the presence of platinum, giving erroneous redox potentials.

Saturated Calomel Electrode (SCE)

This popular reference electrode is easily constructed from readily available materials. A solution is saturated with both calomel (mercurous chloride) and potassium chloride and placed over a layer of mercury which serves as the electrode. A commercial version of this electrode is shown in Figure.

$$Hg \mid Hg_2Cl_2, KC1(sat)$$

The potential can be derived from the primary reaction:

$$Hg_2^{+2} + 2e^- \rightleftharpoons 2Hg$$

And the corresponding Nernst equation:

$$E = E^\circ_{Hg} - 0.059/2 \log 1/\left[Hg_2^{+2}\right]$$

But the concentration of Hg_2^{+2} is in turn controlled by the concentration of Cl^- which is essentially the same as the solubility of KCl. Chloride ion from Hg_2Cl_2 is negligible compared to that from KCl.

$$E = E^{\circ}_{Hg} - 0.059/2 \log \left[Cl^{-2} / K_{Hg^2Cl^2} \right]$$

Where KHg_2Cl_2 is the solubility product of Hg_2Cl_2.

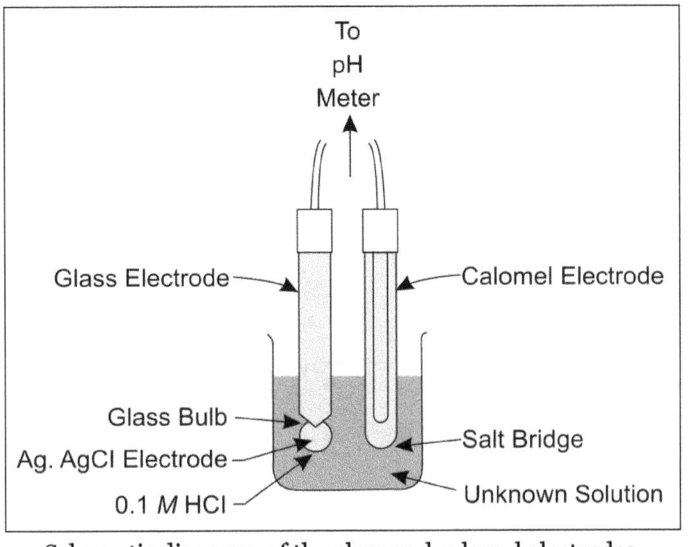

Schematic diagrams of the glass and calomel electrodes.

Rearranging previous equation, we define a new E° term for this electrode reaction:

$$E = E^{\circ}_{Hg} - 0.059/2 \log K_{Hg2Cl2} - 0.059 \log \left[Cl^- \right] = 0.242v.$$

$$E^{\circ}_{Hg2Cl2,Hg}$$

The temperature coefficient of the SCE is practically the same as that of the solubility of KCl; namely,

$$E_{SCE} = +\, 0.242 - 7.6 \times 10^{-4} (t - 25),$$

Where t is the temperature in °C. Because the potential is controlled by a species that does not take part in the electrode reaction, this is known as an electrode of the second order.

Silver Chloride Electrode

A silver wire in contact with chloride ion is readily coated with a layer of AgCl. It is an excellent indicator for chloride ion concentration and a convenient reference electrode.

$$Ag \mid AgCl, Cl^-$$

$$E = E^{\circ}_{Ag+,Ag} - 0.059 \log 1/\left[Ag^+ \right]$$

$$E = E^{\circ}_{Ag+,Ag} + 0.059 \log K_{AgCl} - 0.059 \log \left[Cl^- \right]$$

Glass Electrode

No doubt the most common potentiometric measurement is the determination of hydrogen ion with the glass electrode and a pH meter. Figure is a schematic representation of typical glass and calomel electrodes available commercially. Inside the glass electrode is a silver wire dipping into a solution of ca. 0.1 M HCl. The bulb at the tip of the electrode is made of a special glass composition. The inner surface of this glass membrane contacts the 0.1 M HCl solution and the outer surface contacts the solution in which the pH is to be measured. At each surface the glass membrane absorbs water forming a gel layer. The H^+ ions from the solution can diffuse through the gel layer and replace (ion exchange) sodium or other metal ions in the glass structure. The net result of the diffusion and exchange processes is that a phase boundary potential is set up on each side of the glass membrane with the magnitude determined by the activity of H^+ ion in the contacting solution.

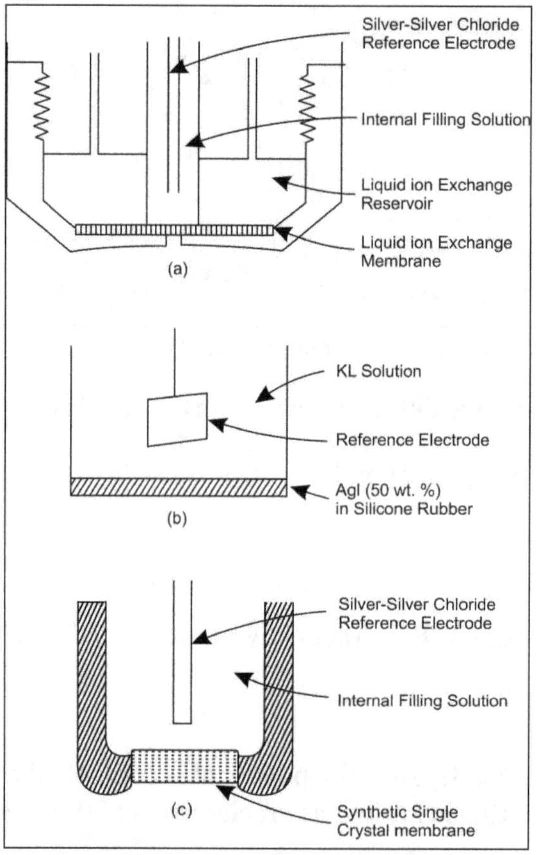

Types of ion-selective electrodes: (a) liquid-liquid membrane; (b) precipitate impregnated membrane; (c) solid state membrane.

The overall potential of the glass electrode consists of several parts: (a) the potential of the internal silver-silver chloride electrode, (b) the potential developed at the inner glass surface, (c) the asymmetry potential (caused by strains and imperfections in the glass membrane) and (d) the potential developed at the outer surface. However, potentials (a), (b), and (c) are all constant for a given electrode, and it is only (d) which changes with pH of the test solution. In practice, it is convenient to combine the constant potential terms and write the potential of the electrode in the usual Nernst form:

$$E = E'_G + 0.059 \log \left[H^+\right]$$

Equation above can be derived from Donnan membrane theory or simply stated as the observed response of the glass electrode.

The glass electrode is a most convenient electrode to use but it has a few drawbacks. The commercial version is moderately expensive and somewhat fragile-the sensitive tip is easily scratched and ruined. The electrical resistance of the glass membrane is exceedingly high (ca. 30 megohms) requiring the use of high impedance amplifiers and special care to prevent electrical leakages in other parts of the circuit. In highly alkaline solution, the concentration of H^+ is so low that other ions in the solution may interfere. For example, in strong NaOH solutions, the electrode may function as a sodium ion indicator electrode to some extent and a correction is necessary for accurate pH measurements. Special glass compositions with low sodium content are available at extra cost. Other special glasses with high sodium content have been developed specifically for measuring Na+ concentrations, and glasses for measuring some other metal ions are on the market.

Ion-Selective Electrodes

The glass electrode just described is one example of a membrane electrode that is more or less selective in response to a single type of ion. We noted that at high pH, or high concentrations of sodium hydroxide, the glass electrode functions partly as a sodium ion electrode. If the composition of the glass is altered, the electrode responds more selectively to sodium ion activity. Several types of membranes are used in a variety of ion-selective electrodes.

These electrodes function to a large extent because of the ion exchange properties of the membrane. By their nature, ion exchange sites are not specific for a given ion. Nevertheless, the sites are somewhat selective and will exhibit a preference for some ions over others, depending primarily on the size and charge of the ions. Selectivities are usually expressed as a selectivity ratio, K, defined as the ratio of concentration of the two ions which will give identical potentiometric response under otherwise identical conditions. As an example, a certain membrane electrode is 100 times more sensitive to Br^- than to Cl^- that is $K_{Br.Cl} = 10^{-2}$ for chloride ion. Then,

$$E = \text{constant} + RT/F \text{ In } \left(a_{Br} + K_{Br.Cl} a_{Cl}\right)$$

$K_{Br,.Cl}$ can also be defined in terms of the potentials measured in two separate solutions, one containing 0.1 M KBr and the other 0.1 M KCl. Then,

$$E_{KCl} = E_{KBr} = RT/F \ln K_{Br,Cl} \times a_{Cl}/a_{Br}$$

Unfortunately, selectivity ratios often vary with the composition of the solution. Therefore they are useful only as a first approximation and are not reliable for making extensive corrections for interfering ions. They are used primarily as an indication of the suitability of an electrode for a particular determination.

Silicone Rubber Matrix Electrodes

Only cations appear to be exchanged at glass membranes. Sparingly soluble salts of the anion to be measured are suitable candidates for anion-selective electrodes. It is not easy to prepare coherent electrode membranes with favorable mechanical properties. Pungor and his coworkers in Hungary have polymerized a mixture of a finely divided silver halide precipitate (grain size 5 to 10 μm) and silicone rubber monomer. When incorporated in these heterogeneous membranes, each silver halide shows a distinct preference for exchanging its own anion with the test solution. Cations are not involved in the potential determining step, therefore the membrane electrode is free from redox interference (strong oxidants or reductants) as is the case with silver metal-silver halide electrodes.

Solid-State Electrodes

Suitably doped single crystals of lanthanum fluoride are used as fluoride electrodes. The detailed composition of the crystal is proprietary information (traces of europium salts are often added). A number of similar electrodes are available for sulfide and the other halides. Selectivity in solid membranes results from the way that ions "move" through the crystal. Ion conduction proceeds through a defect mechanism in which a mobile ion moves into a vacancy in the crystal structure. Thus a new vacancy is created and the charge is propagated through the crystal even though a given ion moves only a small distance. Such vacancies will admit only a certain type of mobile ion.

Liquid-Liquid Electrodes

A large number of highly selective ion exchange materials are available only in the liquid state. However, liquid-liquid electrodes present mechanical problems; for example, the liquid exchanger must be in electrolytic contact with the sample solution with minimal mixing of the two liquid phases. Its viscosity should be high enough to inhibit flow across the interface. The interface between the two liquid phases can be maintained by using a cellulose dialysis membrane which is permeable to ions but not to solvent molecules.

One of the important liquid-liquid electrodes is the calcium electrode. The substance used as a liquid ion exchanger is an organic derivative of phosphoric acid, $(RO)_2POOH$, where R is a long chain organic group. A calcium reference solution contacts the inner surface of the membrane, and the test solution containing calcium contacts the outer surface, thus the measured potential is really the difference between two liquid junction potentials. At the junctions, the organic phosphate group forms a salt selectively with calcium, with selectivity coefficients of: Ca^{+2}/Sr^{+2} 70, Ca^{+2}/Mg^{+2} 200, Ca^{+2}/Ba^{+2} 630, Ca^{+2}/Na^{+} 3200, Ca^{+2}/K^{+} 3200. Similar electrodes are available for potassium and ammonium ions.

Some liquid exchangers are selective because of chelate formation. In $R\text{-}S\text{-}CH_2\text{-}COO^-$, both the sulfur and carboxylate groups participate in chelate formation with metal ions. Copper and lead electrodes are based on this principle.

Because of their selectivity, sensitivity, and rapid, convenient operation, ion-selective electrodes have found many applications. Ions, such as fluoride, are not easily determined by other methods. Previously the determination of fluoride in toothpaste involved steam distillation followed by titration with thorium nitrate. With the fluoride electrode, no separation is necessary and the fluoride can be determined by direct potentiometry. Similarly, the calcium electrode is used extensively in biomedical research. Numerous applications are discussed in Ion-Selective Electrodes, edited by Durst.

Cell Voltage Measurements

Ordinary voltmeters and ammeters require small but appreciable currents to deflect the needle. Since any current flowing through the cell will necessarily change the concentrations of the reacting species, the simple voltmeter is not a suitable device for accurately measuring a cell voltage. Two alternatives are available:

- Electronic Voltmeter: A high input impedance vacuum tube voltmeter requires as little as 10^{-14} A to give an accurately measurable reading. Modern meters are transistorized and require a negligible amount of power from the source to be measured.
- Potentiometer Circuits: In a potentiometer circuit, the voltage to be measured is exactly balanced by a known variable voltage as shown in Figure. The accuracy of determining the null point is limited by the current necessary to detect the off-balance.

The use of a high-impedance voltage amplifier in a potentiometric circuit provides the ultimate accuracy in voltage measurement.

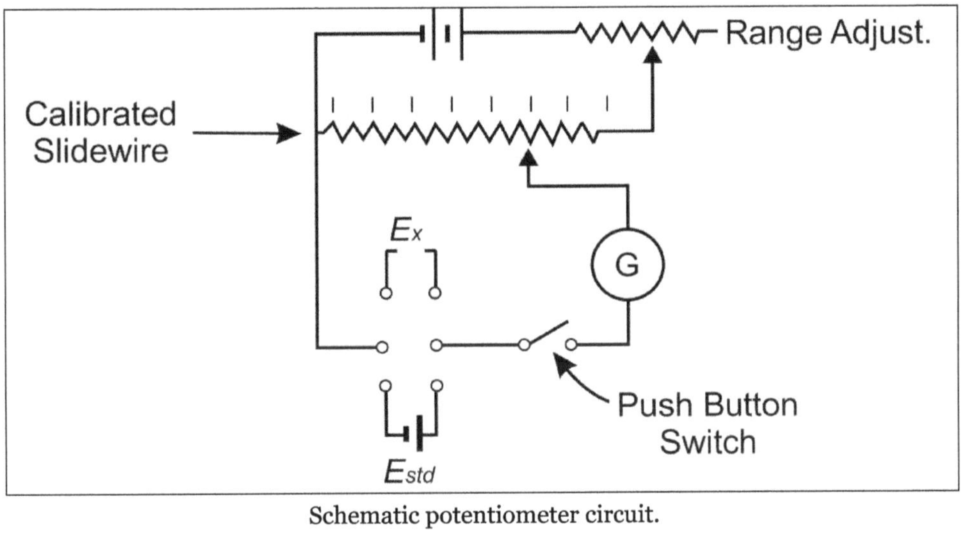

Schematic potentiometer circuit.

pH and the pH Meter

In all of chemistry "pH" is, one of the most used and misused terms. First, let us note the difference between concentration of total titratable hydrogen ion (including that of undissociated acids present), and the concentration of free hydrogen ion. The former can be determined by titration, but the latter cannot because any addition of base will cause further dissociation of any acids present. To determine free hydrogen ion concentration we must make the measurement without disturbing the existing equilibria. The term "pH" refers to free hydrogen ion concentration.

Operational Definition of pH

The definition employed should be capable of theoretical interpretation and practical measurement. Of all the techniques available to measure free hydrogen ion in a solution of unknown composition, none will measure either concentration or activity. The simple definitions, pH = –log

C_H and pH = –log a_H, are therefore not sufficient. Since pH is most often measured with a glass electrode and a pH meter, let us restrict our considerations to this technique. The cell to be used is represented:

Cell 9: Glass | H^+ || SCE

$$E_L = E'_G + 0.059 \log (H^+) \text{ and } E_R = E_{SCE}$$

$$\text{Cell} = E_{SCE} + E_j - E'_G, - 0.059 \log (H^+)$$

Or, combining constants over the bracket,

$$E_{cell} = E^* - 0.059 \log a_H$$
$$= E^* - 0.059 \log C_H f_H$$

In the discussion of activity coefficients, it was shown that it is not possible to measure or compute f_H by itself. However, by convention f_H is equal to $f_\pm$ which can be measured in cells such as Cell 8.

Cell 8: Pt, H_2| H^+, Cl^-, AgCl | Ag

Therefore, for practical purposes, pH is defined as:

$$pH = -\log C_H f_\pm$$

Or in terms of the voltage of either cell 8 or 9 just described,

$$E_{cell} = E^* + 0.059\, pH$$

Or

$$pH = E_{cell} - E^* / 0.059 \text{ at } 25°C$$

Where E* is a different constant for each cell.

Equation above is the operational definition of pH. Thus two approximations are tacitly included in the definition of pH: (a) the junction potential, E_j, remains constant, and (b) $f_H = f_\pm$. The second approximation is reasonably good in dilute aqueous solution, but the first is good only if the value of E_j is small (a few millivolts at most), as it should be with an effective salt bridge.

The pH Scale

The E* term contains so many complications that it can only be determined empirically. A solution of known pH is required. Such a solution has been specified by the National Bureau of Standards using equation $E = constant + RT / F\, In(a_{Br} + K_{Br.Cl} a_{Cl})$ after a careful study of the most nearly ideal systems available. Actually several standard buffers have been specified, because no glass electrode follows the Nernst equation precisely over a wide range of pH values. Thus the working pH scale is an arbitrary set of numbers defined. The pH value is close, but not exactly equal, to -log aH.

When one considers the approximations and experiment-l limitations, it is evident that it is very difficult indeed to reduce the systematic error below ±0.01 pH unit, although small changes in pH can be measured at the ±0.001 level if the random error (electronic noise) can be made quite small.

The pH Meter

The pH meter is basically an electronic voltmeter designed for use with a glass electrode system. From Equation 18-30 we see that 1 pH unit is equivalent to 59 mV at 25 °C. Therefore, the same meter can be used with two scales to read either m V or pH units. The basic circuit diagram of a typical pH meter is given in Figure. The sensitivity of the meter (scale divisions/mY) is adjustable by a temperature control knob to allow for a change in the 0.059 factor with temperature (or to allow for non-ideal response of the electrode). To read absolute mV rather than pH, the shunt used for temperature control is disconnected entirely.

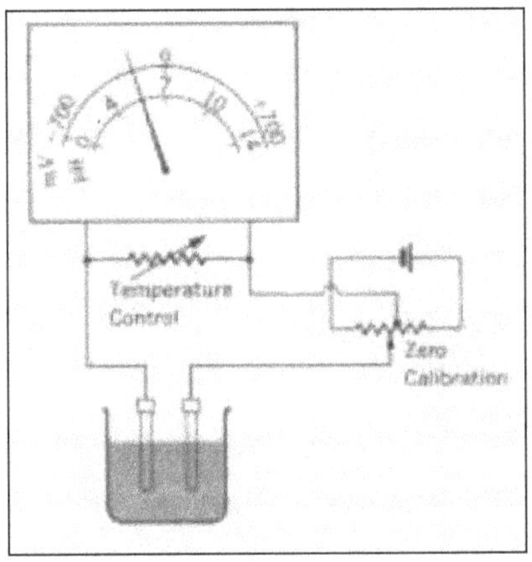

Schematic diagram of a pH meter.

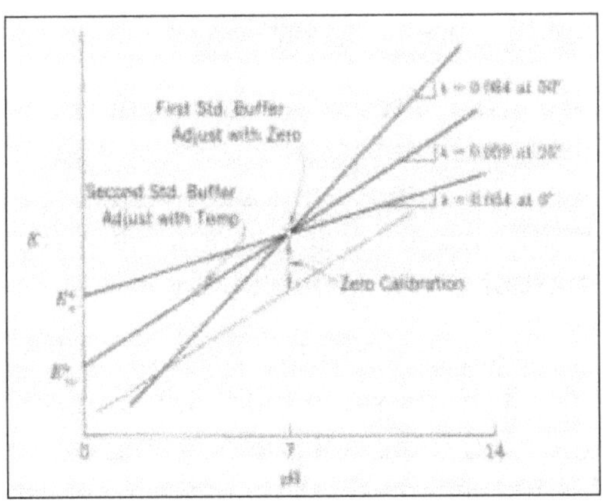

The basic equation of a pH meter. (Not to scale.) Zero calibration raises or lowers response line. Temperature control changes slope of response line.

The zero calibration knob changes the zero setting of the meter by introducing a bias voltage into the meter circuit so that the meter reading will correspond to the pH of the standard buffer. pH meters are often constructed so that the true zero of the voltmeter is at or near pH 7 on the scale. In use, the meter should first be standardized with a pH 7 buffer, adjusting the zero calibration to give the correct reading. Then a second buffer, pH 4 or 10, is substituted and the meter is adjusted to give the correct reading with the temperature control. This procedure establishes the correct linear relationship between mV and pH. At least two points are required to establish the line as shown in Figure, although for rough measurements, one point and the theoretical slope of 0.059 may be adequate for calibration.

Equilibrium Constants

A useful relation is obtained by considering a cell at equilibrium. As we have already observed, the half-cell reactions are always at equilibrium with the corresponding electrode, but the overall cell reaction, in general, is not at equilibrium, unless the cell voltage happens to be zero. Under the latter condition (zero cell voltage) the ratio of activities of products to activities of reactants must satisfy the equilibrium constant, K, for the cell reaction and the following equations pertain:

$$E_{cell} = E^{\circ}_{cell} - 0.059/n \log (\text{products})/(\text{reactants})$$

At equilibrium,

$$E_{cell} = 0 \text{ and } (\text{products})/(\text{reactants}) = K$$

$$0 = E^{\circ}_{cell} - 0.059/n \log K$$

Or

$$\log K = n\ E^{\circ}_{cell} / 0.059 \text{ at } 25\ ^{\circ}\text{C}.$$

Where E°_{cell} and K must correspond to the same reaction. If E°_{cell} is positive, then K is greater than unity, and vice versa. To avoid confusion over signs of E° values, remember that half-reactions (always written as reductions) should be subtracted in a direction to give the required whole reaction, and that E° values corresponding to these half-reactions must be subtracted in the same direction.

In a potentiometric titration the course of the titration reaction is followed by measuring the concentration of one or more of the species potentiometrically. The titration beaker becomes one of the half-cells of an electrochemical cell, along with a convenient reference electrode for the other half-cell. It is not necessary that the titration reaction be a redox reaction as long as one of the species is, or can be made to be, a part of a redox couple.

It is important to distinguish between the titration reaction and the cell reaction. The reactants and products of the titration are all in the same half -cell, and the titration reaction is normally always at equilibrium. On the other hand the cell reaction consists of either of two possible half-reactions in the titration half-cell and the half-reaction in the reference half-cell. The cell reaction is normally not at equilibrium because it is not allowed to take place except for brief periods while measuring the cell voltage.

The Iron(II) Cerium(IV) Titration

Let us study the titration of Fe^{+2} with Ce^{+4}. The experimental features are shown in Figure in which the cell is represented:

$$Hg \mid Hg_2Cl_2, KCl_{(sat)} \mid\mid Fe^{+2}, Fe^{+3}, Ce^{+4}, Ce^{+3} \mid Pt$$

Titration half-Cell: $Fe^{+3} + e^{-} \rightleftharpoons Fe^{+2}$;

$$E = E^{o}_{Fe} - 0.059 \log \left[Fe^{+2}\right] / \left[Fe^{+3}\right]$$

Or

$$Ce^{+4} + e^{-} \rightleftharpoons Ce^{+3};$$

$$E = E^{o}_{Ce} - 0.059 \log \left[Ce^{+3}\right] / \left[Ce^{+4}\right]$$

Reference Half-cell: $Hg_2Cl_2 + 2e^{-} \rightleftharpoons 2Hg + 2Cl^{-}$

Cell reaction: $2Fe^{+3} + 2Hg + 2Cl^{-} \rightleftharpoons 2Fe^{+2} + Hg_2Cl_2$

Or

$$2Ce^{+4} + 2Hg + 2Cl^{-} \rightleftharpoons 2Ce^{+3} + Hg_2Cl_2$$

$$Fe^{+2} + Ce^{+4} \rightleftharpoons Fe^{+3} + Ce^{+3}$$

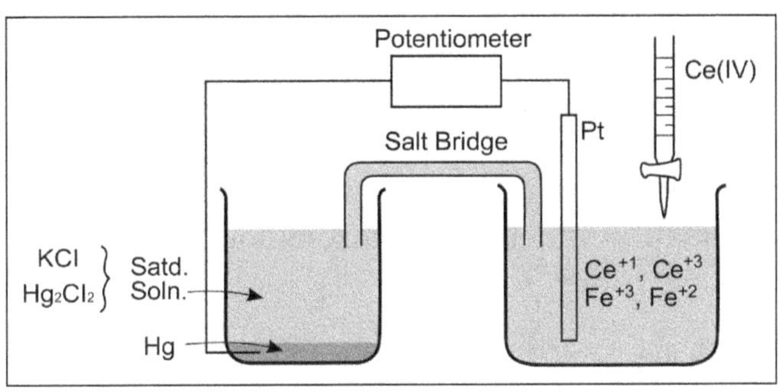

Apparatus for the potentiometric titration of iron(II) with cerium(IV).

Titration Reaction

If we perform the titration, measuring the cell voltage at many points along the way, the data are conveniently plotted as shown in Figure. Now we shall examine the theory that explains this curve and would allow us to predict this curve without ever having to do the experiment. It is convenient to divide the titration into several regions.

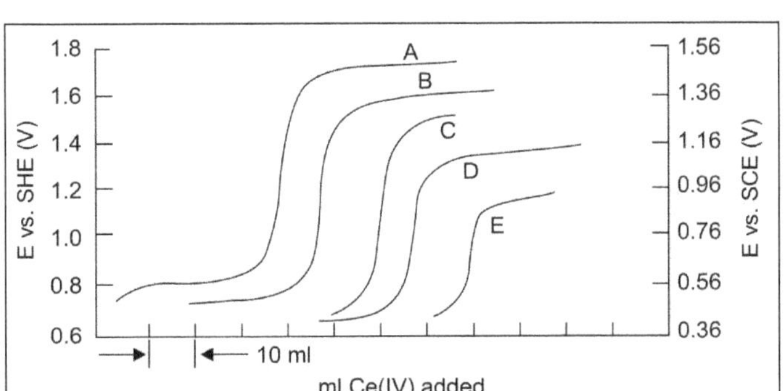

Potentiometric titration curves for iron(II) with cerium(IV) in various acids: A, Theoretical; B, HCIO.; C, HNO$_3$; D, H$_2$SO.: E, HC.

At the beginning the solution contains only Fe^{+2} plus traces of Fe^{+3} due to oxidation by air or hydrogen ion. If the concentration of Fe^{+3} were really zero, the Nernst equation would give us an infinitely negative value for the potential. This would indicate an infinitely powerful reducing agent that would surely reduce water. The initial potential can be measured but not calculated without knowing the extent of this extraneous oxidation. Since the initial potential is of no consequence to our titration, we will ignore the problem.

Once the titration is begun, all four species are present. However, up to the equivalence point, there remains an excess of Fe^{+2} in the half-cell, and the concentration of Ce^{+4} must remain extremely small as a consequence of the very large equilibrium constant for the titration reaction. In this region, the amount of Fe^{+2} converted to Fe^{+3} is directly proportional to the amount of Ce^{+4} added. It is easy to calculate the ratio $[Fe^{+2}]$ J $[Fe^{+3}]$ from the volume of Ce^{+4} added, V, compared to the additional volume increment needed to reach the equivalence point, $V_{ep} - V$.

$$\left[Fe^{+2}\right] / \left[Fe^{+3}\right] = V_{ep} - V / V$$

At the equivalence point, there is no excess of either reactant. By definition, the same number of equivalents of both Fe^{+2} and Ce^{+4} have been added. The large equilibrium constant indicates that each is converted essentially completely to Fe^{+3} and Ce^{+3}, respectively. But the equilibrium constant, though large, is not infinite, and therefore the concentrations of Fe^{+2} and Ce^{+4} cannot be zero. The concentrations of Fe^{+2} and Ce^{+4} could be computed from the value of the equilibrium constant, but there is a more elegant way of calculating the potential without needing to know these concentrations. Remembering that both Nernst equations are valid, we simply add the two together to obtain another equation which is true throughout the titration, but is useful only at the equivalence point:

$$2E = E^{o}_{Fe} + E^{o}_{ce} - 0.059 \log \left[Fe^{+2}\right]\left[Ce^{+3}\right] / \left[Fe^{+3}\right] \left[Ce^{+4}\right]$$

Equation above may resemble a Nernst equation, but it is only a bit of algebraic maneuvering performed with equations $E = E^{o}_{Fe} - 0.059 \log \left[Fe^{+2}\right] / \left[Fe^{+3}\right]$ and $Ce^{+4} + e^{-} \rightleftharpoons Ce^{+3}$; $E = E^{o}_{Ce} - 0.059 \log \left[Ce^{+3}\right] / \left[Ce^{+4}\right]$. We combined the two Nernst equations in this manner only because it leads to a very simple expression for the equivalence point potential. At this point,

$$\left[Fe^{+2}\right] \left[Ce^{+4}\right] \text{ and } \left[Fe^{+3}\right] \left[Ce^{+3}\right]$$

Thus we can cancel out all concentrations in equation:

$2E = E^{o}_{Fe} + E^{o}_{ce} - 0.059 \log \left[Fe^{+2}\right]\left[Ce^{+3}\right] / \left[Fe^{+3}\right] \left[Ce^{+4}\right]$ leaving a value of unity for the concentration ratio and zero for the log term. Therefore,

$$E_{ep} = ½ \left(E^{o}_{Fe} + E^{o}_{Ce}\right)$$

Observe that E_{ep} is independent of concentrations and equal to the arithmetic mean of the two standard potentials. A bit of reflection will show that if n_{ox} for the oxidant and n_{red} for the reductant in the titration reaction are not the same, we must modify our treatment. The two log terms of the individual Nernst equations cannot be combined unless the factor preceding the log is the same for both. The two factors can be made the same if we multiply the Nernst equation for the oxidant by n_{ox} and multiply that for the reductant by n_{red}. Thus the E_{ep} will become a weighted mean:

$$E_{ep} = n_{ox}E^{o}_{ox} + n_{red}E^{o}_{red}/n_{red} + n_{ox}$$

For some un symmetric titration reactions, for example, those involving H^{+} or complexing agents, not all concentrations will cancel. In these cases, E_{ep} will depend on the concentration of such species.

Beyond the equivalence point, an excess of Ce^{+4} has been added. The amount of the excess is clearly that which has been added after the equivalence point has been reached, and its concentration can be easily calculated from the total volume of the solution. Likewise, the amount of Ce^{+3} in the solution can be calculated from the amount of Ce^{+4} required to reach the equivalence point-essentially all of it was reduced to Ce^{+3} by the Fe^{+2} in the original solution,

$$\left[Ce^{+3}\right]/\left[Ce^{+4}\right] = V_{ep}/V - V_{ep}$$

The theoretical titration curve is plotted in figure along with several actual titration curves obtained experimentally in different solutions. There are two important reasons why the actual curves differ from the theoretical curve. All four of the species involved in this titration readily form complexes and are subject to hydrolysis and dimerization. Furthermore, highly charged ions have activity coefficients which deviate greatly from unity. A different theoretical curve would have been obtained if we had used the appropriate formal potentials. A second cause of the difference between theoretical and actual curves is the irreversible behavior of many electrodes.

On the other hand, theoretical curves do approximate the actual curves in many important respects, and are useful in obtaining a graphical picture of what happens during a titration. The shape of a titration curve in the vicinity of the equivalence point, and in particular the slope of the curve at that point, tells us how sharp the end point will be, and within what range of potential we must stop in order to obtain a given accuracy.

Redox Indicator

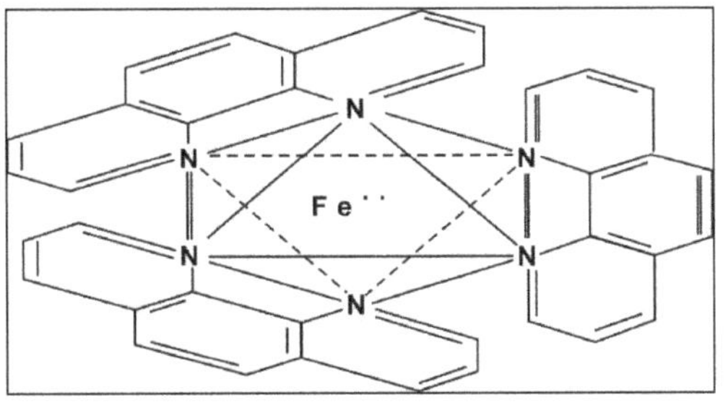

It is not always necessary to set up the potentiometric equipment and actually measure an electrode potential in order to determine the end point of a titration. A redox indicator may be used to give a visual indication that this point has been reached. The color change takes place upon oxidation or reduction of the indicator itself. If the standard potential of the indicator system is close to the equivalence point potential of the titration reaction, the indicator will exist half in each form at the end point and display a color intermediate to the two forms.

Ferroin is a very stable, intensely red complex of 1,10-phenanthroline with Fe(II): or $(Phen)_2Fe^{+2}$ when oxidized, the ferric complex is faintly blue:

$$(Phen)_3\,Fe^{+3} + e^- \rightleftharpoons (Phen)_3\,Fe^{+2};\quad E' = 1.06\ V \text{ in } 0.5\ M\ H_2SO_4$$

$$E' = 1.06 - 0.059 \log \left(\text{red form}\right) / \left(\text{blue form}\right)$$

With the Nernst equation, it is easy to show that this indicator exists 90% in the red form at E = 1.00 V, 50% in each form at E = 1.06 V, and 90% in the blue form at E = 1.12 V. The transition range is sufficiently close to the equivalence point potential (1.19 V) of the ferrous-eerie titration so that the end-point error is less than 0.1%. Most redox indicators exhibit complicated reaction mechanisms and do not exactly follow the Nernst equation.

Polarography

In analytic chemistry, Polarography is also called polarographic analysis or voltammetry which is an electrochemical method of analyzing solutions of reducible or oxidizable substances. In general, polarography is a technique in which the electric potential (or voltage) is varied in a regular manner between two sets of electrodes (indicator and reference) while the current is monitored. The shape of a polarogram depends on the method of analysis selected, the type of indicator electrode used, and the potential ramp that is applied. The Figure shows five selected methods of polarography; the potential ramps are applied to a mercury indicator electrode, and the shapes of the resulting polarograms are compared.

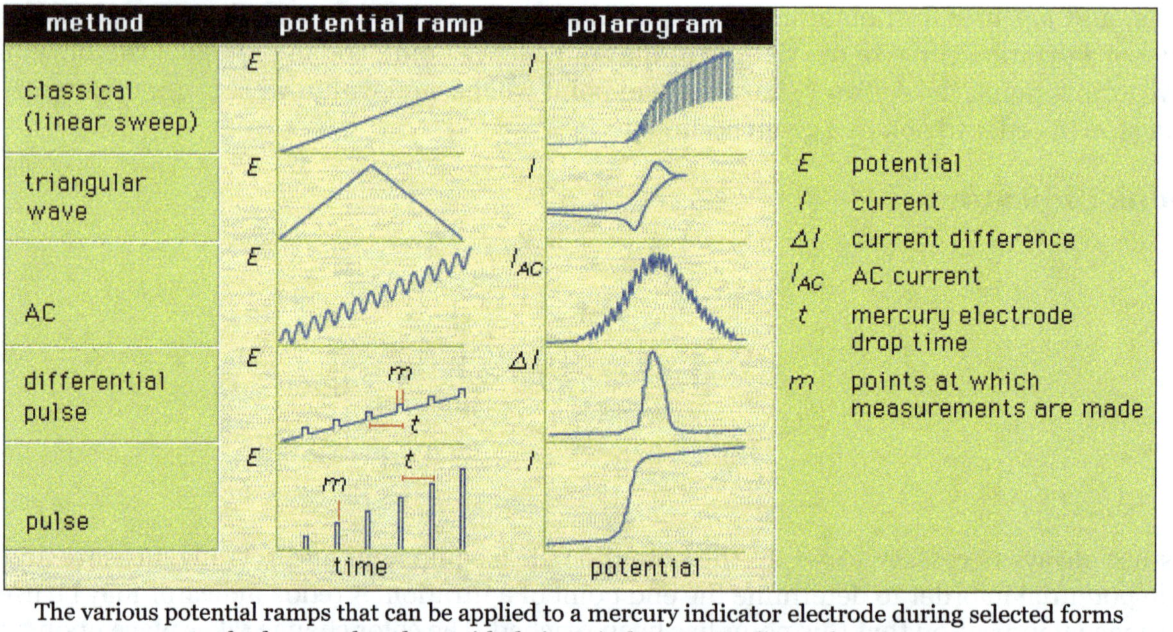

The various potential ramps that can be applied to a mercury indicator electrode during selected forms of polarography, along with their typical corresponding polarograms.

The majority of the chemical elements can be identified by polarographic analysis, and the method is applicable to the analysis of alloys and to various inorganic compounds. Polarography is also used to identify numerous types of organic compounds and to study chemical equilibria and rates of reactions in solutions.

The solution to be analyzed is placed in a glass cell containing two electrodes. One electrode consists of a glass capillary tube from which mercury slowly flows in drops, and the other is commonly a pool of mercury. The cell is connected in series with a galvanometer (for measuring the flow of current) in an electrical circuit that contains a battery or other source of direct current and a device for varying the voltage applied to the electrodes from zero up to about two volts. With the dropping mercury electrode connected (usually) to the negative side of the polarizing voltage, the voltage is increased by small increments, and the corresponding current is observed on the galvanometer. The current is very small until the applied voltage is increased to a value large enough to cause the substance being determined to be reduced at the dropping mercury electrode. The current increases rapidly at first as the applied voltage is increased above this critical value but gradually attains a limiting value and remains more or less constant as the voltage is increased further. The critical voltage required to cause the rapid increase in current is characteristic of, and also serves to identify, the substance that is being reduced (qualitative analysis). Under proper conditions the constant limiting current is governed by the rates of diffusion of the reducible substance up to the surface of the mercury drops, and its magnitude constitutes a measure of the concentration of the reducible substance (quantitative analysis). Limiting currents also result from the oxidation of certain oxidizable substances when the dropping electrode is the anode.

When the solution contains several substances that are reduced or oxidized at different voltages, the current-voltage curve shows a separate current increase (polarographic wave) and limiting current for each. The method is thus useful in detecting and determining several substances simultaneously and is applicable to relatively small concentrations—e.g., 10^{-6} up to about 0.01 mole per litre, or approximately 1 to 1,000 parts per 1,000,000.

Electrodeposition

Electrodeposition is an electroanalytical technique in which an element is quantitatively deposited on an electrode. Electrodeposition, a short version of electrolytic deposition, is the conventional process of coating a thin layer of one metal on top of a different metal to modify its surface properties. Based on the principle of electrolysis, it is a process that uses electrical current to reduce the cations of a desired material from an electrolyte and coat those materials as a thin film onto a conductive substrate surface. This is done to achieve the desired electrical and corrosion resistance, reduce wear and friction, improve heat tolerance and for decoration. Hence the surface properties of a material can be tailored and new materials for a variety of applications can be explored through this technique.

Electrolysis

For potentiometry, the cell is not allowed to operate. With zero current, it is assumed that the electrode is in equilibrium with the solution and that concentrations do not change with time. Such cells, however, possess an e.m.f. whether or not they are operating, and are sources of electrical energy if the circuit is completed analytical current allowed to flow. With a finite current, the concentrations must change as the cell reaction proceeds. The e.m.f. gradually approaches zero as the concentrations approach the equilibrium state. Now let us add another component to the circuit. The cell in Figure is connected to an external battery or power supply, E_{appl}, so that the two electro motive forces compete with each other. If E_{appl}, is larger than E_{cell}, then the current is reversed. Similarly, the cell reaction is upturned from its usual spontaneous direction. In place of copper metal dissolving into the solution, copper is placed on the electrode surface. The normal spontaneous cell reaction for this cell is:

Pt Electrode: $O_2 + 4H^+ + 4e^- \rightleftharpoons 2H_2 0$

Cu Electrode: $Cu^{+2} + 2e^- \rightleftharpoons Cu$

Cell: $2Cu + O_2 + 4H^+ \rightleftharpoons 2Cu^{+2} + 2H_2 0$

With corresponding potentials:

$$E_{Pt} = E^o_{O_2,H_2O} - \frac{0.059}{4}\log\frac{1}{Po_2\left[H^+\right]^4} = 1.23 - \frac{0.059}{4}\log\frac{1}{(0.2)(0.2)^4} = 1.18 \text{ V}$$

$$E_{Cu} = E^o_{Cu^{+2},Cu} - \frac{0.059}{2}\log\frac{1}{\left[Cu^{+2}\right]} = 0.34 - \frac{0.059}{2}\log\frac{1}{0.1} = 0.31\text{V}$$

$$E_{cell} = E_{Pt} - E_{Cu} = 1.18 - 0.31 = 0.87V$$

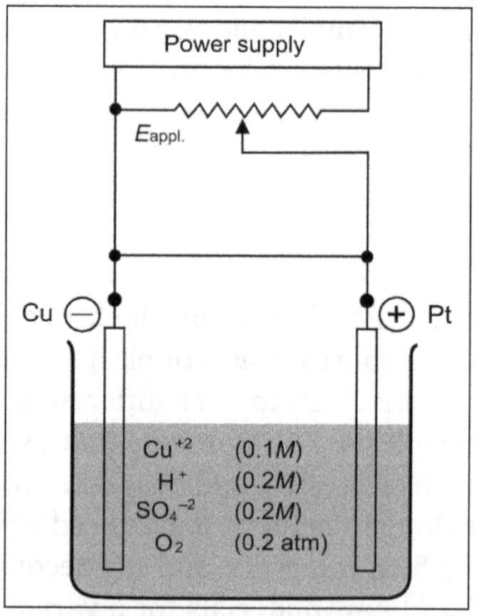

Apparatus for the electrodeposition of copper.

The cell voltage is often called the "back e.m.f." In actual cases, an E_{app1} slightly greater than E_{cell} is required to drive the cell reaction backwards. The extra voltage, called the overvoltage, is usually caused by the irreversibility of the electrode reactions. The basic equation for electrolysis is:

$$E_{appl} = E_{cell} + E_{ov} + iR$$

In which all quantities are considered to be positive. E_{ov} is the sum of the overvoltages (if any) at the electrodes, i represent the current, and R represents the total resistance of the circuit. With current flowing, the concentrations change with time which results in changes in all three terms on the right side of equation above. There are many kinds of electrolysis which make use of equation above. One or more of the terms in this equation may be held constant giving rise to several special techniques.

If the electrolysis of copper just described is continued for any length of time, the concentration of cupric ion in the electrolyte must decrease. In fact, copper ion consumed at the cathode is being replaced by hydrogen ion produced in the anode reaction. The net result is that the cell voltage must increase; therefore, the applied voltage must be increased in order to maintain the current flow. If the current is continued long enough to be sure that the cell reaction is essentially completed, the original concentration of copper ion can be determined from the increase in weight of the electrode. This has been the classical way to determine copper and a few other metals, but it finds little use in organic analysis.

Controlled Current Electrolysis

If E_{appl} is continuously adjusted to maintain a constant current, and if conditions are such that only a single electrode reaction takes place at the electrode (100% current efficiency), then it is not necessary to weigh the electrode. The amount of the reaction can be readily determined from Faraday's law:

$$F \times \text{equiv reacted} = i \times t$$

Where t is the time in seconds and F is the number of coulombs per equivalent (96493) or the Faraday constant. However, we have noted in the electrolysis of copper, that the cathode potential must be made more negative with time in order to continue the reduction of copper ion. Concurrently the hydrogen ion concentration is increasing, and eventually the cathode potential will be sufficiently negative to reduce hydrogen ion as well as copper ion. This is a typical situation that is to say; it is rarely possible to maintain a single electrode reaction. The method, as such, finds little use, but is the basis for coulometric titrations to be described shortly.

Controlled Cathode (or Anode) Potential Electrolysis

If a (third) auxiliary reference electrode is placed within the electrolysis cell and connected through a voltage measuring device, then the potential of (either) one of the working electrodes can be monitored, as shown in Figure. The E_{appl} is then adjusted throughout the electrolysis to maintain a constant cathode or anode potential. In this way, we can be fairly sure that only a single reaction is taking place and that the number of coulombs passed is directly related to the substance to be determined. However, the current decreases with time and approaches zero asymptotically. Therefore, Equation $F \times \text{equiv reacted} = i \times t$ must be modified:

$$F \times \text{equiv reacted} = \int_0^t i \, dt$$

In addition to analytical applications, controlled potential electrolysis is useful in identifying reaction products and the number of electrons taking part in a reaction. (These are not always obvious in complex organic reactions). The method also serves as a preparative tool. The electrode acts as an oxidant or reductants with a constant known potential of any value one desires. Some preparations for which no suitable reagents are available can be easily carried out.

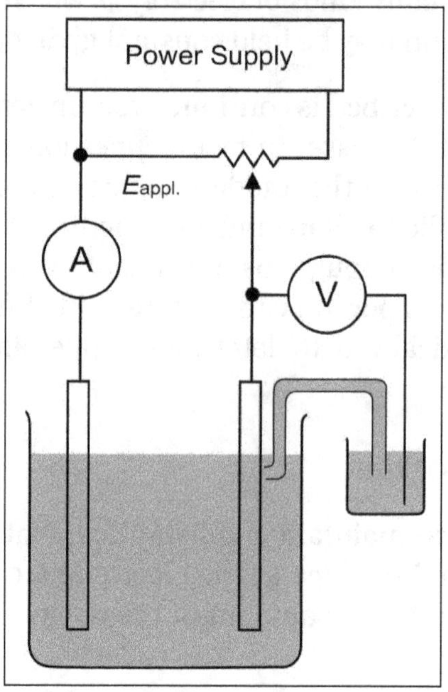

Apparatus for electrolysis with controlled cathode (or anode) potential. Signal from voltmeter may be used to adjust E_{appl} automatically as required.

Secondary Coulometric Titrations

We have seen that with controlled cathode (or anode) potential electrolysis it is relatively easy to maintain a single electrode reaction, or 100% current efficiency. However, the current necessarily decreases with time in a logarithmic fashion so that the area under the i versus t curve must be found by integration. Constant current electrolysis allows an easy determination of coulombs, i x t, but since the electrode potential necessarily changes with time, other electrode reactions must occur. Therefore we cannot maintain 100% current efficiency with respect to the desired reaction.

Consider the electrolytic oxidation of ferrous ion at a platinum anode. If the current is kept constant, the extent of reaction is directly proportional to the time of electrolysis, but the electrode potential must increase and another reaction will eventually take over. On the other hand, if the potential is kept constant, only the desired reaction will occur, but the rate of reaction will decrease exponentially.

The advantages of both these techniques are nearly combined in a secondary coulometric titration. In this method the substance to be determined is indirectly "titrated" with a reagent that is electrolytically generated in situ. Instead of oxidizing ferrous ion at the electrode, as in the above examples, let us add a large excess of cerous ion prior to the electrolysis. When the current is started, the ferrous ion in the vicinity of the anode is exhausted. The potential tends to increase rapidly, but is prevented from doing so because of the large amount of cerous ion available for oxidation. The ceric ion so produced is immediately mixed with the solution and reacts with the remaining ferrous ion. The concentration of ceric ion stays at the trace level until the entire ferrous ion is oxidized. The end point of the ferrous reaction can be monitored by a visual (redox) indicator, by potentiometry using an extra set of electrodes and a separate circuit, or by amperometry. Thus it is possible to maintain a reasonably constant electrode potential for the duration of the electrolysis that is carried out at constant current.

A coulometric titration set up includes a beaker for titration and two independent circuits as shown in figure. The generator circuit requires a source of constant current (which should be adjustable over a range of values) and a pair of electrodes. For precision and convenience, the switch controlling the current source also controls a timer. The detector circuit requires a suitable indicator electrode, reference electrode, and potentiometer. The signal from the potentiometer can be used to stop the titration and the timer. For repetitive determinations, the timer can be calibrated in grams, equivalents, or even % composition of a particular weight of sample.

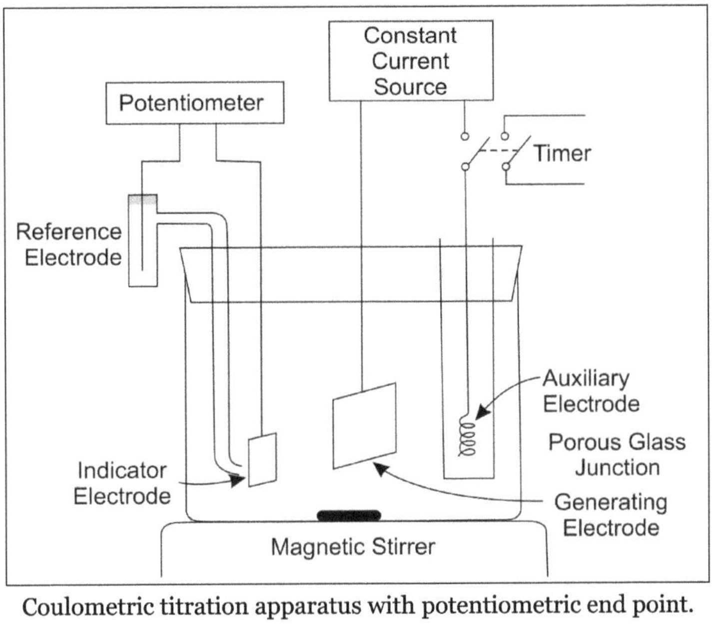

Coulometric titration apparatus with potentiometric end point.

Many types of titrants normally used in volumetric analysis include oxidation, reduction, acid-base, precipitants, complexing agents, and several metal ions. There is no need to prepare or store standard solutions. Thus, chlorine and chromous ion are as easy to use as sodium chloride.

Chemical primary standards are replaced by electrical measurements, or, stated another way; the electron becomes the universal primary standard. Small currents and short times can be measured accurately. In principle, the accuracy is limited by the uncertainty in the value of the Faraday. In practice, the accuracy is limited by the reproducibility of the end point.

Acid-base titrations are readily performed by coulometry. Both acid and base can be generated either internally within the solution, or externally as shown in Figure. The external apparatus is particularly convenient for generating small amounts of carbonate-free base as needed. A large number of organic compounds react with the halogens that are easily generated from the corresponding halide at a platinum electrode. The determinations of 8-hydroxyquinoline with electrolytically generated bromine and the bromine number of unsaturated compounds are two examples.

The determination of traces of pesticide residues in foods is a difficult but important analytical problem. The sensitivity of most analytical methods is not adequate, and even with the best gas chromatographs an ultra-sensitive, specific detector is required. Many common pesticides contain chlorine or sulfur which are readily converted to HCl and H_2S in reactors built into the end of the chromatographic column. After selective absorption of these gases in an electrolytic cell, the coulometric titrator with electrolytic generation of silver ion is an ideal detector for this problem.

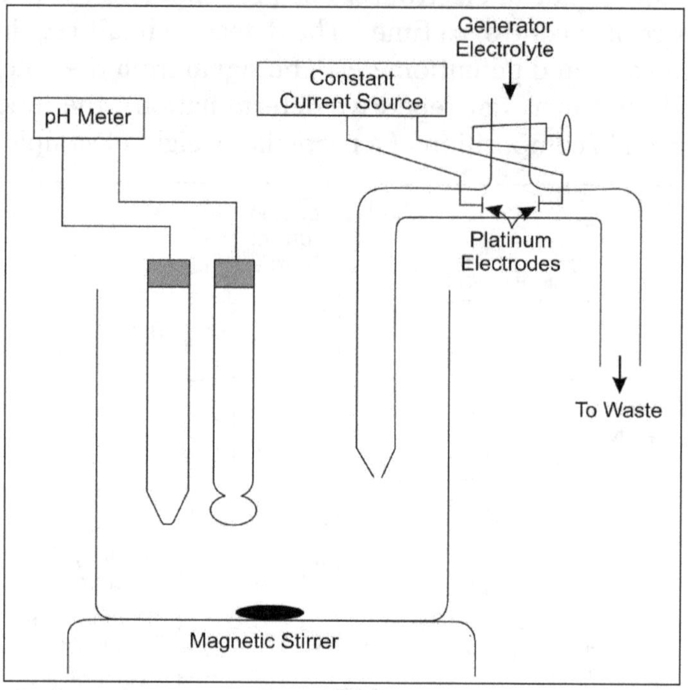

Coulometric apparatus for external generation of acid or base.

The advantages of coulometric titration can be summarized as follows:

- Small amounts of substances can be determined accurately.
- No standard solutions are required.
- Unstable intermediates can be used as titrants.
- The apparatus required is inexpensive and easily automated.

6

Common Separation Techniques

There are various techniques which are used for separation such as chromatography, extraction, filtration, decantation, sublimation, evaporation, simple distillation, fractional distillation, centrifugation and magnetic separation. The diverse applications of these chemical separation techniques have been thoroughly discussed in this chapter.

Separation Process

Separation techniques are those techniques that can be used to separate two different states of matter such as liquid and solid. Such separation techniques include filtration or evaporation. Separation process, or a separation method, or simply a separation, is a methodology to attain any mass transfer phenomenon that converts a mixture of substances into two or more distinct product mixtures.

Separations Based on Size

Size is the simplest physical property we can exploit in a separation. To accomplish the separation we use a porous medium through which only the analyte or the interferent can pass. Examples of size-based separations include filtration, dialysis, and size-exclusion. In a filtration we separate a particulate interferent from dissolved analytes using a filter whose pore size retains the interferent. The solution passing through the filter is called the filtrate, and the material retained by the filter is the retentate. Gravity filtration and suction filtration using filter paper are techniques with which you should already be familiar. Membrane filters, available in a variety of micrometer pores sizes, are the method of choice for particulates that are too small to be retained by filter paper. Figure provides information about three types of membrane filters.

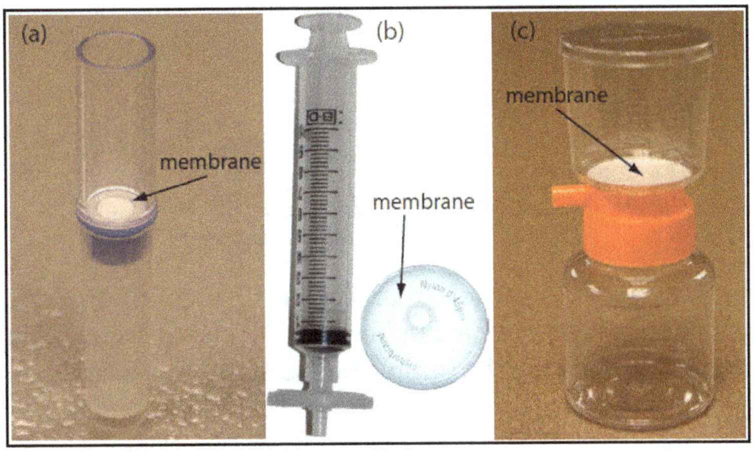

In the figure above, three types of membrane filters for separating analytes from interferents. (a) A centrifugal filter for concentrating and desalting macromolecular solutions. The membrane has a nominal molecular weight cut-off of 1 × 106 g/mol. The sample is placed in the upper reservoir and the unit is placed in a centrifuge. Upon spinning the unit at 2000×g–5000×g, the filtrate collects in the bottom reservoir and the retentate remains in the upper reservoir. (b) A 0.45 μm membrane syringe filter. The photo on the right shows the membrane filter in its casing. In the photo on the left, the filter is attached to a syringe. Samples are placed in the syringe and pushed through the filter. The filtrate is collected in a test tube or other suitable container. (c) A disposable filter system with a 0.22 μm cellulose acetate membrane filter. The sample is added to the upper unit and vacuum suction is used to draw the filtrate through the membrane and into the lower unit. To store the filtrate, the top half of the unit is removed and a cap placed on the lower unit. The filter unit shown here has a capacity of 150 mL.

Dialysis is another example of a separation technique that uses size to separate the analyte and the interferent. A dialysis membrane is usually constructed from cellulose and fashioned into tubing, bags, or cassettes. Figure shows an example of a commercially available dialysis cassette. The sample is injected into the dialysis membrane, which is sealed tightly by a gasket, and the unit is placed in a container filled with a solution whose composition is different from the sample. If the concentration of a particular species is different on the membrane's two sides, the resulting concentration gradient provides a driving force for its diffusion across the membrane. While small species freely pass through the membrane, larger species are unable to pass. Dialysis is frequently used to purify proteins, hormones, and enzymes. During kidney dialysis, metabolic waste products, such as urea, uric acid, and creatinine, are removed from blood by passing it over a dialysis membrane.

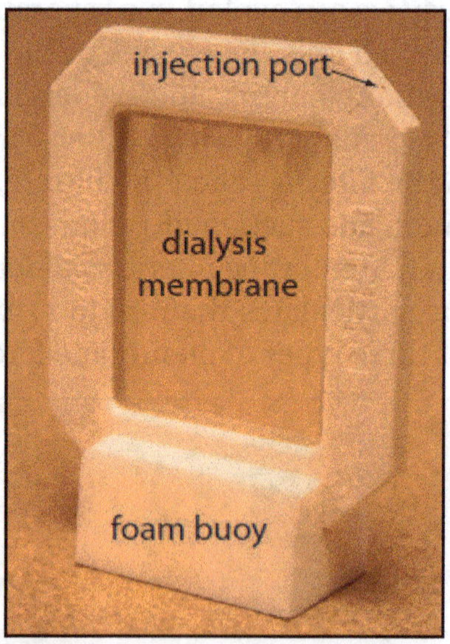

Figure: Example of a dialysis cassette.

The dialysis membrane in this unit has a molecular weight cut-off of 10 000 g/mol. Two sheets of the membrane are separated by a gasket and held in place by the plastic frame. Four ports, one of which is labeled, provide a means for injecting the sample between the dialysis membranes. The cassette is inverted and submerged in a beaker containing the external solution, which is stirred using a stir bar. A foam buoy, used as a stand in the photo, serves as a float so that the unit remains suspended above the stir bar. The external solution is usually replaced several time during dialysis. When dialysis is complete, the solution remaining in the cassette is removed through an injection port.

Size-exclusion chromatography is a third example of a separation technique that uses size as a means for effecting a separation. In this technique a column is packed with small, approximately 10-μm, porous polymer beads of cross-linked dextrin or polyacrylamide. The pore size of the particles is controlled by the degree of cross-linking, with more cross-linking producing smaller pore sizes. The sample is placed into a stream of solvent that is pumped through the column at a fixed flow rate. Those species too large to enter the pores pass through the column at the same rate as the solvent. Species entering into the pores take longer to pass through the column, with smaller species needing more time to pass through the column. Size-exclusion chromatography is widely used in the analysis of polymers, and in biochemistry, where it is used for the separation of proteins.

Separations Based on Mass or Density

If the analyte and the interferent differ in mass or density, then a separation using centrifugation may be possible. The sample is placed in a centrifuge tube and spun at a high angular velocity, measured in revolutions per minute (rpm). The sample's constituents experience a centrifugal force that pulls them toward the bottom of the centrifuge tube. The species experiencing the greatest centrifugal force has the fastest sedimentation rate, and is the first to reach the bottom of the centrifuge tube. If two species have equal density, their separation is based on mass, with the heavier species having the greater sedimentation rate. If the species are of equal mass, then the species with the largest density has the greatest sedimentation rate.

Centrifugation is an important separation technique in biochemistry. For example, lists conditions for separating selected cellular components. We can separate lysosomes from other cellular components by several differential centrifugations, in which we divide the sample into a solid residue and a supernatant solution. After destroying the cells, the solution is centrifuged at 15 000 $\times g$ (a centrifugal force that is 15 000 times that of the earth's gravitational force) for 20 minutes, leaving a solid residue of cell membranes and mitochondria. The supernatant, which contains the lysosomes, is isolated by decanting it from the residue and centrifuged at 30 000 $\times g$ for 30 minutes, leaving a solid residue of lysosomes. Figure shows a typical centrifuge capable of produce the centrifugal forces needed for biochemical separations.

Table: Conditions for Separating Selected Cellular Components by Centrifugation

Components	Centrifugal Force ($\times g$)	Time (min)
eukaryotic cells	1000	5
cell membranes, nuclei	4000	10
mitochondria, bacterial cells	15000	20
lysosomes, bacterial membranes	30000	30
ribosomes	100000	180

Figure: *Bench-top centrifuge capable of reaching speeds up to 14 000 rpm and centrifugal forces of 20 800 × g. This particular centrifuge is refrigerated, allowing samples to be cooled to temperatures as low as −4° C.*

An alternative approach to differential centrifugation is density gradient centrifugation. To prepare a sucrose density gradient, for example, a solution with a smaller concentration of sucrose—and, thus, of lower density—is gently layered upon a solution with a higher concentration of sucrose. Repeating this process several times, fills the centrifuge tube with a multi-layer density gradient. The sample is placed on top of the density gradient and centrifuged using a force greater than 150 000 × *g*. During centrifugation, each of the sample's components moves through the gradient until it reaches a position where its density matches the surrounding sucrose solution. Each component is isolated as a separate band positioned where its density is equal to that of the local density within the gradient. Figure below provides an example of a typical sucrose density centrifugation for separating plant thylakoid membranes.

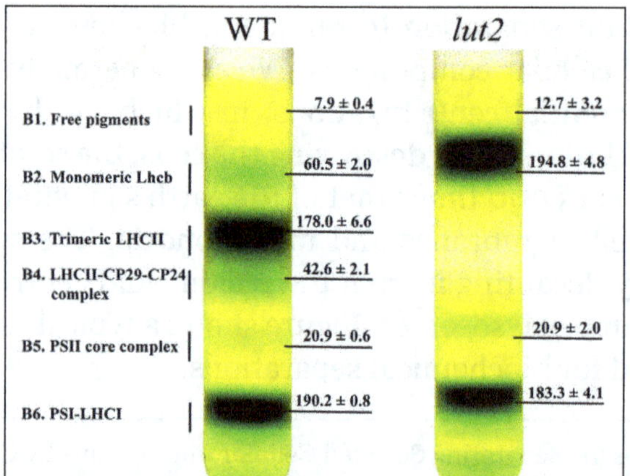

Figure: Example of a sucrose density gradient centrifugation of thylakoid membranes from wild type (WT) and lut2 plants. The thylakoid membranes were extracted from the plant's leaves and separated by centrifuging in a 0.1–1 M sucrose gradient for 22 h at 280 000 × g at 4° C. Six bands and their chlorophyll contents are shown.

Separations based on Complexation Reactions

One widely used technique for preventing an interference is to bind the interferent in a strong, soluble complex that prevents it from interfering in the analyte's determination. This process is known as masking. As shown in Table, a wide variety of ions and molecules are useful masking agents, and, as a result, selectivity is usually not a problem.

Table: Selected Inorganic and Organic Masking Agents for Metal Ions	
Masking Agent	Elements Whose Ions Can Be Masked
CN^-	Ag, Au, Cd, Co, Cu, Fe, Hg, Mn, Ni, Pd, Pt, Zn
SCN^-	Ag, Cd, Co, Cu, Fe, Ni, Pd, Pt, Zn
NH_3	Ag, Co, Ni, Cu, Zn
F^-	Al, Co, Cr, Mg, Mn, Sn, Zn
$S_2O_3^{2-}$	Au, Ce, Co, Cu, Fe, Hg, Mn, Pb, Pd, Pt, Sb, Sn, Zn
tartrate	Al, Ba, Bi, Ca, Ce, Co, Cr, Cu, Fe, Hg, Mn, Pb, Pd, Pt, Sb, Sn, Zn
oxalate	Al, Fe, Mg, Mn
thioglycolic acid	Cu, Fe, Sn

Separations Based on a Change of State

Because an analyte and its interferent are usually in the same phase, we can achieve a separation if one of them undergoes a change in its physical state or its chemical state.

Changes in Physical State

When the analyte and the interferent are miscible liquids, a distillation may be possible if their boiling points are significantly different. Figure below shows the progress of a distillation as a plot of temperature versus the mixture's vapor-phase and liquid-phase composition. The initial mixture (point A), contains more interferent than analyte. When this solution is brought to its boiling point, the vapor phase in equilibrium with the liquid phase is enriched in analyte (point B). The horizontal line connecting points A and B represents this vaporization equilibrium. Condensing the vapor phase at point B, by lowering the temperature, creates a new liquid phase whose composition is identical to that of the vapor phase (point C). The vertical line connecting points B and C represents this condensation equilibrium. The liquid phase at point C has a lower boiling point than the original mixture, and is in equilibrium with the vapor phase at point D. This process of repeated vaporization and condensation gradually separates the analyte and interferent.

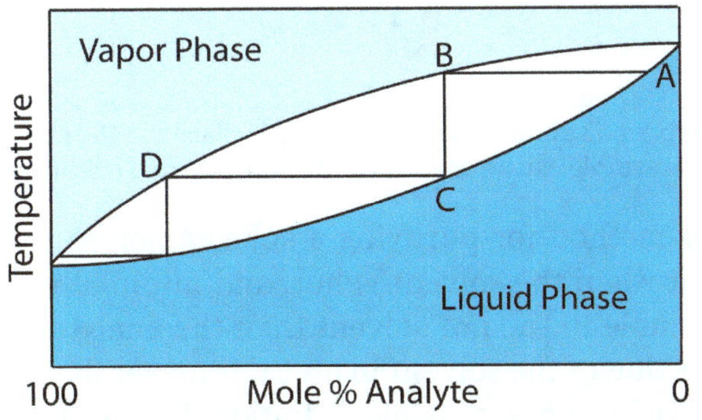

Figure: Boiling point versus composition diagram for a near-ideal solution consisting of a low-boiling analyte and a high-boiling interferent. The horizontal lines represent vaporization equilibria and the vertical lines represent condensation equilibria.

Two experimental set-ups for distillations are shown in figure. The simple distillation apparatus shown in Figure (a) does not produce a very efficient separation, and is useful only for separating a volatile analyte (or interferent) from a non-volatile interferent (or analyte), or for an analyte and interferent whose boiling points differ by more than 150° C. A more efficient separation is achieved by a fractional distillation. Packing the distillation column with a high surface area material, such as a steel sponge or glass beads, provides more opportunity for the repeated process of vaporization and condensation necessary to effect a complete separation.

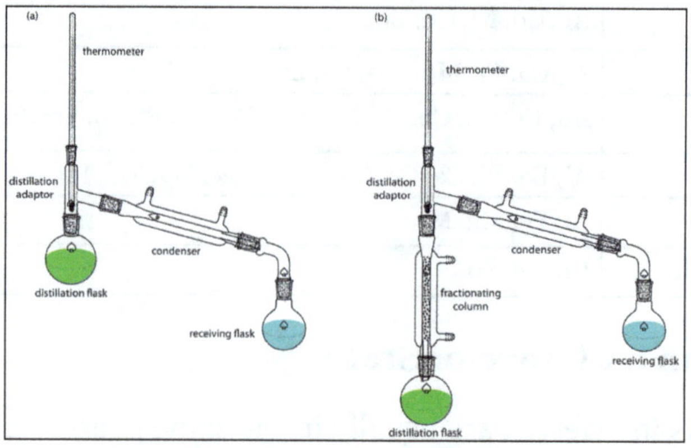

Figure: Typical experimental set-up for (a) a simple distillation, and (b) a fractional distillation.

When the sample is a solid, sublimation may provide a useful separation of the analyte and the interferent. The sample is heated at a temperature and pressure below the analyte's triple point, allowing it to vaporize without passing through the liquid state. Condensing the vapor recovers the purified analyte. A useful analytical example of sublimation is the isolation of amino acids from fossil mollusk shells and deep-sea sediments.

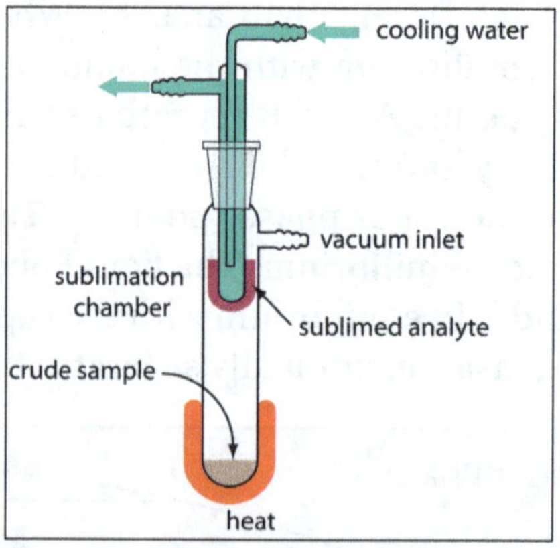

Figure: Typical experimental setup for a sublimation. The sample is placed in the sublimation chamber, which may be evacuated. Heating the sample causes the analyte to vaporize and sublime onto a cooled probe.

Recrystallization is another method for purifying a solid. A solvent is chosen in which the analyte's solubility is significant when the solvent is hot, and minimal when the solvent is cold. The interferents must be less soluble in the hot solvent than the analyte, or present in much smaller amounts. After heating a portion of the solvent in an Erlenmeyer flask, small amounts of the sample are added until undissolved sample is visible. Additional hot solvent is added until the sample redissolves, or until only insoluble impurities remain. This process of adding sample and solvent is repeated until the entire sample has been added to the Erlenmeyer flask. Any insoluble impurities are removed by filtering the hot solution. The solution is allowed to cool slowly, promoting the growth of large, pure crystals, and then cooled in an ice bath to minimize solubility losses. The purified sample is isolated by filtration and rinsed to remove any soluble impurities. Finally, the sample is dried to remove any remaining traces of the solvent. Further purification, if necessary, can be accomplished by additional recrystallizations.

Changes in Chemical State

Distillation, sublimation, and recrystallization use a change in physical state as a means of separation. Chemical reactivity also can be a useful tool for separating analytes and interferents. For example, we can separate SiO_2 from a sample by reacting it with HF to form SiF_4. Because SiF_4 is volatile, it is easy to remove by evaporation. If we wish to collect the reaction's volatile product, then a distillation is possible. For example, we can isolate the NH_4^+ in a sample by making the solution basic and converting it to NH_3. The ammonia is then removed by distillation. Table provides additional examples of this approach for isolating inorganic ions.

Table: Examples of Using a Chemical Reaction and a Distillation to Separate an Inorganic Analyte From Interferents

Analyte	Treatment	Isolated Species
CO_3^{2-}	$CO_3^{2-}(aq) + 2H_3O^+(aq) \rightarrow CO_2(g) + 3H_2O(l)$	CO_2
NH_4^+	$NH_4^+(aq) + OH^-(aq) \rightarrow NH_3(g) + H_2O(l)$	NH_3
SO_3^{2-}	$SO_3^{2-}(aq) + 2H_3O^+(aq) \rightarrow SO_2(g) + 3H_2O(l)$	SO_2
S^{2-}	$S^{2-}(aq) + 2H_3O+(aq) \rightarrow H_2S(g) + 2H_2O(l)$	H_2S

Another reaction for separating analytes and interferents is precipitation. Two important examples of using a precipitation reaction in a separation are the pH-dependent solubility of metal oxides and hydroxides, and the solubility of metal sulfides.

Separations based on the pH-dependent solubility of oxides and hydroxides usually use strong acids, strong bases, or an NH_3/NH_4Cl buffer. Most metal oxides and hydroxides are soluble in hot concentrated HNO_3, although a few oxides, such as WO_3, SiO_2, and SnO_2 remain insoluble even under these harsh conditions. In determining the amount of Cu in brass, for example, we can avoid an interference from Sn by dissolving the sample with a strong acid and filtering to remove the solid residue of SnO_2.

Most metals form a hydroxide precipitate in the presence of concentrated NaOH. Those metals forming amphoteric hydroxides, however, do not precipitate because they react to form higher-order hydroxo-complexes. For example, Zn^{2+} and Al^{3+} do not precipitate in concentrated NaOH because the form the soluble complexes $Zn(OH)_3^-$ and $Al(OH)_4^-$. The solubility of Al^{3+} in concentrated NaOH allows us to isolate aluminum from impure samples of bauxite, an ore of Al_2O_3. After crushing the ore, we place it in a solution of concentrated NaOH, dissolving the Al_2O_3 and forming $Al(OH)_4^-$. Other oxides in the ore, such as Fe_2O_3 and SiO_2, remain insoluble. After filtering, we recover the aluminum as a precipitate of $Al(OH)_3$ by neutralizing some of the OH^- with acid.

The pH of an NH_3/NH_4Cl buffer (pK_a = 9.26) is sufficient to precipitate most metals as the hydroxide. The alkaline earths and alkaline metals, however, do not precipitate at this pH. In addition, metal ions that form soluble complexes with NH_3, such as Cu^{2+}, Zn^{2+}, Ni^{2+}, and Co^{2+} also do not precipitate under these conditions.

The use of S^{2-} as a precipitating reagent is one of the earliest examples of a separation technique. In Fresenius's 1881 text *A System of Instruction in Quantitative Chemical Analysis*, sulfide is frequently used to separate metal ions from the remainder of the sample's matrix.[14] Sulfide is a useful reagent for separating metal ions for two reasons:

(1) Most metal ions, except for the alkaline earths and alkaline metals, form insoluble sulfides; and

(2) These metal sulfides show a substantial variation in solubility. Because the concentration of S^{2-} is pH-dependent, we can control which metal precipitates by adjusting the pH. For example, in Fresenius's gravimetric procedure for determining Ni in ore samples, sulfide is used three times as a means of separating Co^{2+} and Ni^{2+} from Cu^{2+} and, to a lesser extent, from Pb^{2+}.

Separations Based on a Partitioning Between Phases

The most important group of separation techniques uses a selective partitioning of the analyte or interferent between two immiscible phases. If we bring a phase containing a solute, S, into contact with a second phase, the solute partitions itself between the two phases, as shown by the following equilibrium reaction.

$$S_{phase\,1} \rightleftharpoons S_{phase\,2}$$

The equilibrium constant for reaction $S_{phase\,1} \rightleftharpoons S_{phase\,2}$

$$K_D = \frac{[S_{phase\,2}]}{[S_{phase\,1}]}$$

is called the distribution constant or partition coefficient. If K_D is sufficiently large, then the solute moves from phase 1 to phase 2. The solute remains in phase 1 if the partition coefficient is sufficiently small. When we bring a phase containing two solutes into contact with a second phase, if K_D is favorable for only one of the solutes a separation of the solutes is possible. The physical states of the phases are identified when describing the separation process, with the phase containing the sample listed first. For example, if the sample is in a liquid phase and the second phase is a solid, then the separation involves liquid–solid partitioning.

Extraction between Two Phases

We call the process of moving a species from one phase to another phase an extraction. Simple extractions are particularly useful for separations where only one component has a favorable partition coefficient. Several important separation techniques are based on a simple extraction, including liquid–liquid, liquid–solid, solid–liquid, and gas–solid extractions.

Liquid–Liquid Extractions

A liquid–liquid extraction is usually accomplished with a separatory funnel. After placing the two liquids in the separatory funnel, we shake the funnel to increase the surface area between the phases. When the extraction is complete, we allow the liquids to separate. The stopcock at the bottom of the separatory funnel allows us to remove the two phases.

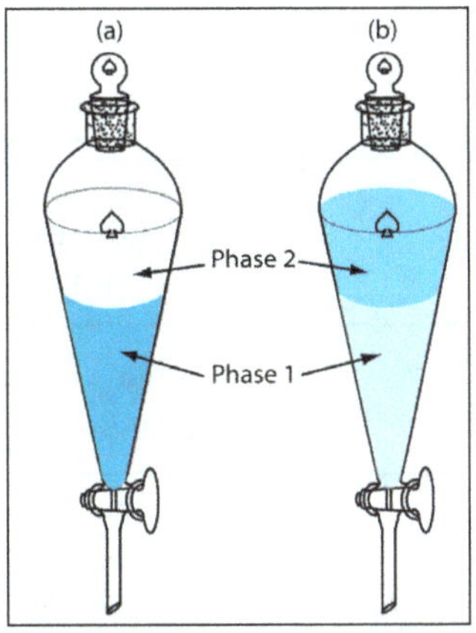

Figure: Example of a liquid–liquid extraction using a separatory funnel. (a) Before the extraction, 100% of the analyte is in phase 1. (b) After the extraction, most of the analyte is in phase 2, although some analyte remains in phase 1. Although one liquid–liquid extraction can result in the complete transfer of analyte, a single extraction usually is not sufficient.

We also can carry out a liquid–liquid extraction without a separatory funnel by adding the extracting solvent to the sample container. Pesticides in water, for example, are preserved in the field by extracting them into a small volume of hexane. Liquid–liquid microextractions, in which the extracting phase is a 1-mL drop suspended from a microsyringe.

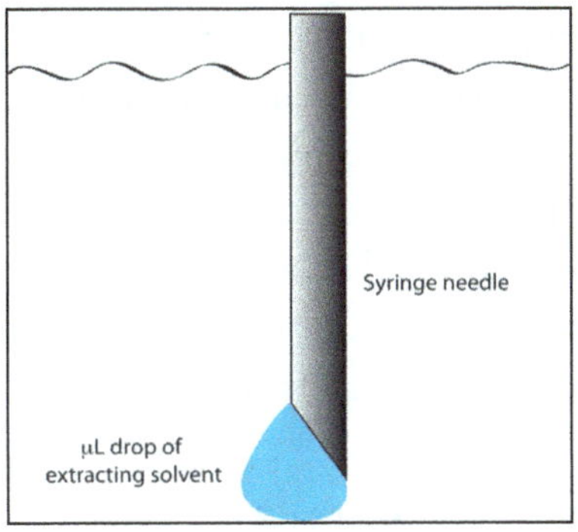

Figure: *Schematic of a liquid–liquid* microextraction showing a syringe needle with a μL drop of the extracting solvent.

Solid Phase Extractions

In a solid phase extraction of a liquid sample, we pass the sample through a cartridge containing a solid adsorbent, several examples of which are shown in figure. The choice of adsorbent is determined by the species we wish to separate. Table provides several representative examples of solid adsorbents and their applications.

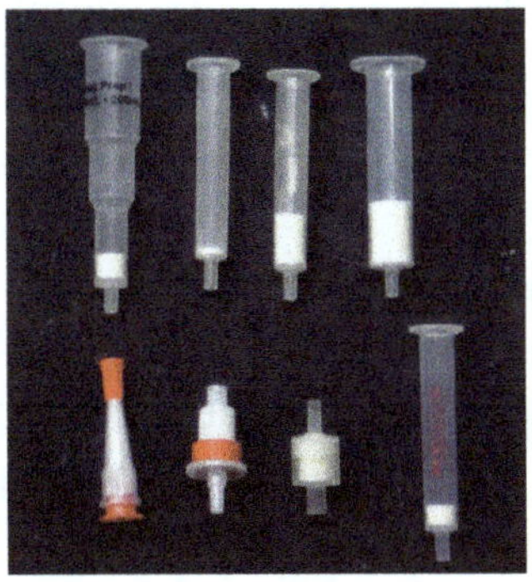

Table: Representative Adsorbents for the Solid Phase Extraction of Liquid Samples

Adsorbent	Structure	Properties and Uses
silica	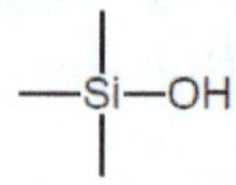	• retains low to moderate polarity species from organic matricies • fat soluble vitamins, steroids
aminopropyl	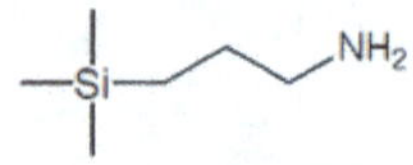	• retains polar compounds • carbohydrates, organic acids
cyanopropyl		• retains wide variety of species from aqueous and organic matricies • pesticides, hydrophobic peptides
diol	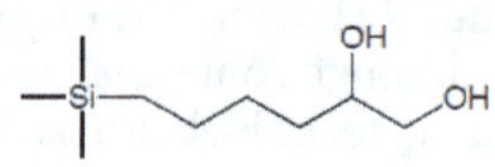	• retains wide variety of species from aqueous and organic matricies • proteins, peptides, fungicides
octadecyl (C-18)	$-C_{18}H_{37}$	• retains hydrophobic species from aqueous matricies • caffeine, sedatives, polyaromatic hydrocarbons, carbohydrates, pesticides
octyl (C-8)	$-C_8H_{17}$	• similar to C-18

As an example, let's examine a procedure for isolating the sedatives secobarbital and phenobarbital from serum samples using a C-18 solid adsorbent. Before adding the sample, the solid phase cartridge is rinsed with 6 mL each of methanol and water. Next, a 500-μL sample of serum is pulled through the cartridge, with the sedatives and matrix interferents retained by a liquid–solid extraction. Washing the cartridge with distilled water removes any interferents. Finally, we elute the sedatives using 500 μL of acetone. In comparison to a liquid–liquid extraction, a solid phase extraction has the advantage of being easier, faster, and requiring less solvent.

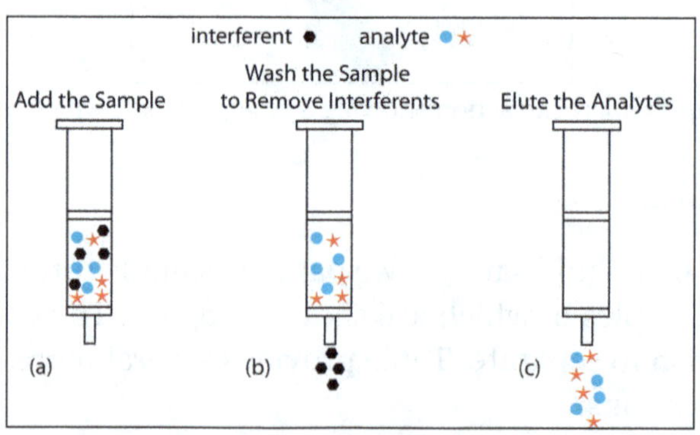

Figure: Steps in a typical solid phase extraction. After preconditioning the solid phase cartridge with solvent, (a) the sample is added to the cartridge, (b) the sample is washed to remove interferents, and (c) the analytes eluted.

Continuous Extractions

An extraction is possible even if the analyte has an unfavorable partition coefficient, provided that the sample's other components have significantly smaller partition coefficients. Because the analyte's partition coefficient is unfavorable, a single extraction will not recover all the analyte. Instead we continuously pass the extracting phase through the sample until a quantitative extraction is achieved.

A continuous extraction of a solid sample is carried out using a Soxhlet extractor. The extracting solvent is placed in the lower reservoir and heated to its boiling point. Solvent in the vapor phase moves upwards through the tube on the far right side of the apparatus, reaching the condenser where it condenses back to the liquid state. The solvent then passes through the sample, which is held in a porous cellulose filter thimble, collecting in the upper reservoir. When the solvent in the upper reservoir reaches the return tube's upper bend, the solvent and extracted analyte are siphoned back to the lower reservoir. Over time the analyte's concentration in the lower reservoir increases.

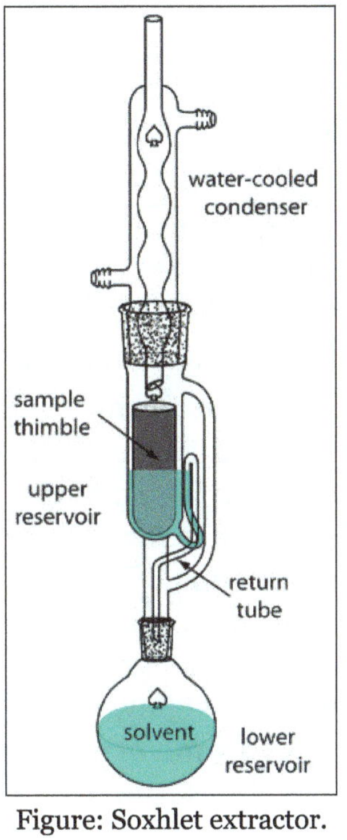

Figure: Soxhlet extractor.

Microwave-assisted extractions have replaced Soxhlet extractions in some applications. The process is the same as that described earlier for a microwave digestion. After placing the sample and the solvent in a sealed digestion vessel, a microwave oven is used to heat the mixture. Using a sealed digestion vessel allows the extraction to take place at a higher temperature and pressure, reducing the amount of time needed for a quantitative extraction. In a Soxhlet extraction the temperature is limited by the solvent's boiling point at atmospheric pressure. When acetone is the solvent, for example, a Soxhlet extraction is limited to 56° C, but a microwave extraction can reach 150° C.

Two other continuous extractions deserve mention. Volatile organic compounds (VOCs) can be quantitatively removed from liquid samples by a liquid–gas extraction. As shown in figure, an inert purging gas, such as He, is passed through the sample. The purge gas removes the VOCs, which are swept to a primary trap where they collect on a solid absorbent. A second trap provides a means for checking to see if the primary trap's capacity is exceeded. When the extraction is complete, the VOCs are removed from the primary trap by rapidly heating the tube while flushing with He. This technique is known as a purge-and-trap. Because the analyte's recovery may not be reproducible, an internal standard is necessary for quantitative work.

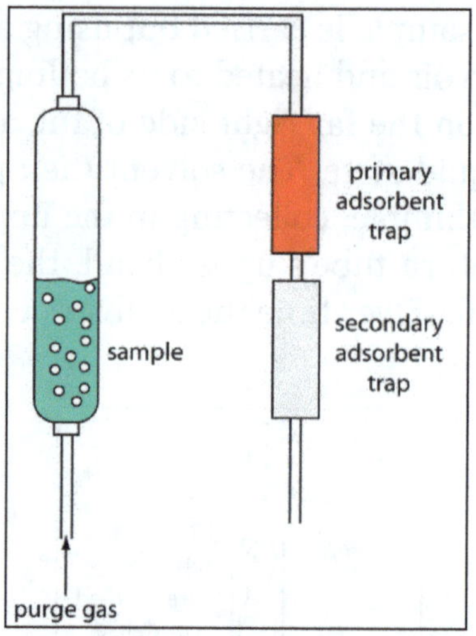

Figure: Schematic diagram of a purge-and-trap system for extracting volatile analytes. The purge gas releases the analytes, which subsequently collect in the primary adsorbent trap. The secondary adsorption trap is monitored for evidence of the analyte's breakthrough.

Continuous extractions also can be accomplished using supercritical fluids. If we heat a substance above its critical temperature and pressure, it forms a supercritical fluid whose properties are between those of a gas and a liquid. A supercritical fluid is a better solvent than a gas, which makes it a better reagent for extractions. In addition, a supercritical fluid's viscosity is significantly less than that of a liquid, making it easier to push it through a particulate sample. One example of a supercritical fluid extraction is the determination of total petroleum hydrocarbons (TPHs) in soils, sediments, and sludges using supercritical CO_2. An approximately 3-g sample is placed in a 10-mL stainless steel cartridge and supercritical CO_2 at a pressure of 340 atm and a temperature of 80°C is passed through the cartridge for 30 minutes at flow rate of 1–2 mL/min. To collect the TPHs, the effluent from the cartridge is passed through 3 mL of tetrachloroethylene at room temperature. At this temperature the CO_2 reverts to the gas phase and is released to the atmosphere.

Chromatographic Separations

In an extraction, the sample is one phase and we extract the analyte or the interferent into a second phase. We also can separate the analyte and interferents by continuously passing one sample-free phase, called the mobile phase, over a second sample-free phase that remains fixed or stationary. The sample is injected into the mobile phase and the sample's components partition themselves between the mobile phase and the stationary phase. Those components with larger partition coefficients are more likely to move into the stationary phase, taking a longer time to pass through the system. This is the basis of all chromatographic separations. Chromatography provides both a separation of analytes and interferents, and a means for performing a qualitative or quantitative analysis for the analyte.

Chromatography as a Separation Process

Most of us have got our papers wet at some time or another, but have you ever noticed what happens to the ink as the water spreads? It doesn't always smudge and blur, as you might expect.

Sometimes it splits up into weird colored streaks that creep across the page. When that happens, you're seeing chromatography in action. In this case it's totally accidental, but we can also use it by design to split up mixtures and other substances into their components. Chromatography is actually one of the most useful analytical techniques chemists have at their disposal, helpful in everything from identifying biological materials to finding clues at crime scenes.

Chromatography is based on the principle where molecules in mixture applied onto the surface or into the solid, and fluid stationary phase (stable phase) is separating from each other while moving with the aid of a mobile phase. The factors effective on this separation process include molecular characteristics related to adsorption (liquid-solid), partition (liquid-solid), and affinity or differences among their molecular weights. Because of these differences, some components of the mixture stay longer in the stationary phase, and they move slowly in the chromatography system, while others pass rapidly into mobile phase, and leave the system faster.

Based on this approach three components form the basis of the chromatography technique.

- Stationary phase: This phase is always composed of a "solid" phase or "a layer of a liquid adsorbed on the surface a solid support".
- Mobile phase: This phase is always composed of "liquid" or a "gaseous component."
- Separated molecules.

The type of interaction between stationary phase, mobile phase, and substances contained in the mixture is the basic component effective on separation of molecules from each other. Chromatography methods based on partition are very effective on separation, and identification of small molecules as amino acids, carbohydrates, and fatty acids. However, affinity chromatographies (ie. ion-exchange chromatography) are more effective in the separation of macromolecules as nucleic acids, and proteins. Paper chromatography is used in the separation of proteins, and in studies related to protein synthesis; gas-liquid chromatography is utilized in the separation of alcohol, esther, lipid, and amino groups, and observation of enzymatic interactions, while molecular-sieve chromatography is employed especially for the determination of molecular weights of proteins. Agarose-gel chromatography is used for the purification of RNA, DNA particles, and viruses.

Stationary phase in chromatography, is a solid phase or a liquid phase coated on the surface of a solid phase. Mobile phase flowing over the stationary phase is a gaseous or liquid phase. If mobile phase is liquid it is termed as liquid chromatography (LC), and if it is gas then it is called gas chromatography (GC). Gas chromatography is applied for gases, and mixtures of volatile liquids, and solid material. Liquid chromatography is used especially for thermal unstable, and non-volatile samples.

The purpose of applying chromatography which is used as a method of quantitative analysis apart from its separation, is to achive a satisfactory separation within a suitable timeinterval. Various chromatography methods have been developed to that end. Some of them include column chromatography, thin-layer chromatography (TLC), paper chromatography, gas chromatography, ion exchange chromatography, gel permeation chromatography, high-pressure liquid chromatography, and affinity chromatography.

Column Chromatography

Since proteins have difference characteristic features as size, shape, net charge, stationary phase used, and binding capacity, each one of these characteristic components can be purified using chromatographic methods. Among these methods, most frequently column chromatography is applied. This technique is used for the purification of biomolecules. On a column (stationary phase) firstly the sample to be separated, then wash buffer (mobile phase) are applied. Their flow through inside column material placed on a fiberglass support is ensured. The samples are accumulated at the bottom of the device in a time-dependent and volume-dependent manner.

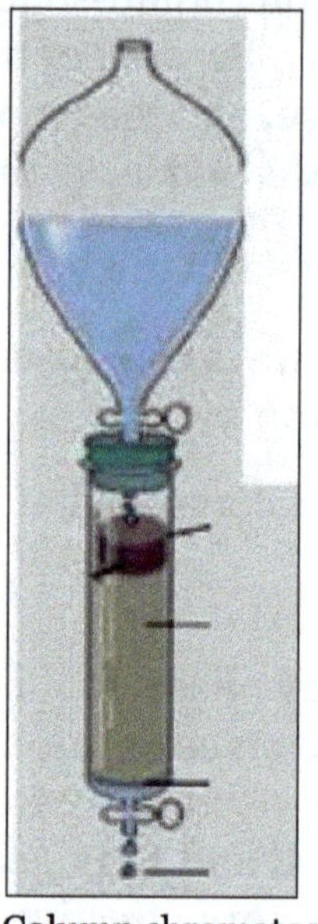

Figure: Column chromatography.

Ion-exchange Chromatography

Ion-exchange chromatography is based on electrostatic interactions between charged protein groups, and solid support material (matrix). Matrix has an ion load opposite to that of the protein to be separated, and the affinity of the protein to the column is achieved with ionic ties. Proteins are separated from the column either by changing pH, concentration of ion salts or ionic strength of the buffer solution. Positively charged ion-exchange matrices are called anion-exchange matrices, and adsorb negatively charged proteins. While matrices bound with negatively charged groups are known as cation-exchange matrices, and adsorb positively charged proteins.

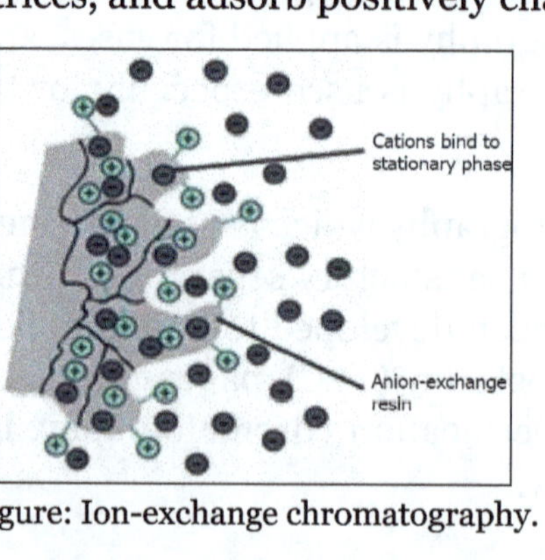

Figure: Ion-exchange chromatography.

Gel- permeation (Molecular Sieve) Chromatography

The basic principle of this method is to use dextran containing materials to separate macromolecules based on their differences in molecular sizes. This procedure is basically used to determine molecular weights of proteins, and to decrease salt concentrations of protein solutions. In gel-permeation column stationary phase consists of inert molecules with small pores. The solution containing molecules of different dimensions are passed continuously with a constant flow rate through the column. Molecules larger than pores cannot permeate into gel particles, and they are retained between particles within a restricted area. Larger molecules pass through spaces between porous particles, and move rapidly through inside the column. Molecules smaller than the pores are diffused into pores, and as molecules get smaller, they leave the column with proportionally longer retention times. Sephadeks G type is the most frequently used column material. Besides, dextran, agorose, polyacrylamide are also used as column materials.

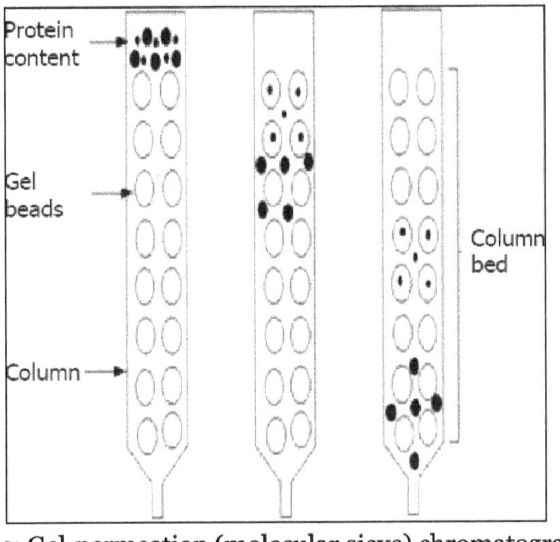

Figure: Gel-permeation (molecular sieve) chromatography.

Affinity Chromatography

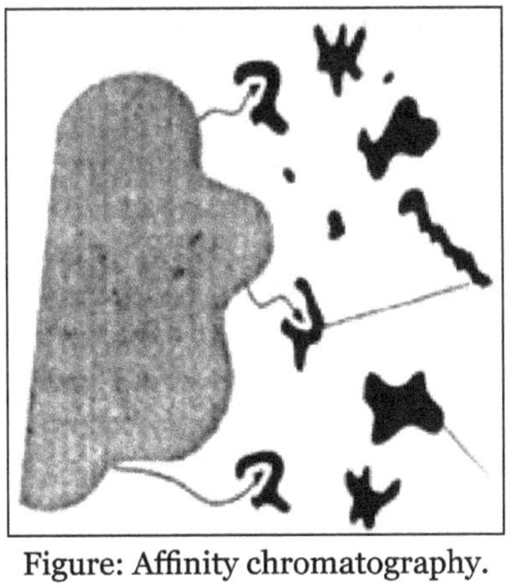

Figure: Affinity chromatography.

This chromatography technique is used for the purification of enzymes, hormones, antibodies, nucleic acids, and specific proteins. A ligand which can make a complex with specific protein (dextran, polyacrylamide, cellulose etc.) binds the filling material of the column. The specific protein which makes a complex with the ligand is attached to the solid support (matrix), and retained in the column, while free proteins leave the column. Then the bound protein leaves the column by means of changing its ionic strength through alteration of pH or addition of a salt solution.

Paper Chromatography

In paper chromatography support material consists of a layer of cellulose highly saturated with water. In this method a thick filter paper comprised the support, and water drops settled in its pores made up the stationary "liquid phase." Mobile phase consists of an appropriate fluid placed in a developing tank. Paper chromatography is a "liquid-liquid" chromatography.

Thin-layer Chromatography

Thin-layer chromatography is a "solid-liquid adsorption" chromatography. In this method stationary phase is a solid adsorbent substance coated on glass plates. As adsorbent material all solid substances used. in column chromatography (alumina, silica gel, cellulose) can be utilized. In this method, the mobile phase travels upward through the stationary phase. The solvent travels up the thin plate soaked with the solvent by means of capillary action. During this procedure, it also drives the mixture priorly dropped on the lower parts of the plate with a pipette upwards with different flow rates. Thus the separation of analytes is achieved. This upward travelling rate depends on the polarity of the material, solid phase, and of the solvent.

In cases where molecules of the sample are colorless, florescence, radioactivity or a specific chemical substance can be used to produce a visible colored reactive product so as to identify their positions on the chromatogram. Formation of a visible color can be observed under room light or UV light. The position of each molecule in the mixture can be measured by calculating the ratio between the distances travelled by the molecule and the solvent. This measurement value is called relative mobility, and expressed with a symbol R_f. R_f value is used for qualitative description of the molecules.

Gas Chromatography

In this method stationary phase is a column which is placed in the device, and contains a liquid stationary phase which is adsorbed onto the surface of an inert solid. Gas chromatography is a "gas-liquid" chromatography. Its carrier phase consists of gases as He or N_2. Mobile phase which is an inert gas is passed through a column under high pressure. The sample to be analyzed is vaporized, and enters into a gaseous mobile phase phase. The components contained in the sample are dispersed between mobile phase and stationary phase on the solid support. Gas chromatography is a simple, multifaceted, highly sensitive, and rapidly applied technique for the extremely excellent separation of very minute molecules. It is used in the separation of very little amounts of analytes.

Dye-ligand Chromatography

Development of this technique was based on the demonstration of the ability of many enzymes to bind purine nucleotides for Cibacron Blue F3GA dye. The planar ring structure with negatively charged groups is analogous to the structure of NAD. This analogy has been evidenced by demonstration of the binding of Cibacron Blue F3GA dye to adenine, ribose binding sites of NAD. The dye behaves as an analogue of ADP-ribose. The binding capacity of this type adsorbents is 10–20-fold stronger than that of the affinity of other adsorbents. Under appropriate pH conditions, elution with high-ionic strength solutions and using ion-exchange property of adsorbent, the adsorbed proteins are separated from the column.

Hydrophobic Interaction Chromatography

In this method the adsorbents prepared as column material for the ligand binding in affinity chromatography are used. HIC technique is based on hydrophobic interactions between side chains bound to chromatography matrix.

Pseudoaffinity Chromatography

Some compounds as anthraquinone dyes, and azo-dyes can be used as ligands because of their affinity especially for dehydrogenases, kinases, transferases, and reeducates The mostly known type of this kind of chromatography is immobilized metal affinity chromatography (IMAC).

High-pressure Liquid Chromatography

Using this chromatography technique it is possible to perform structural, and functional analysis, and purification of many molecules within a short time, This technique yields perfect results in the separation, and identification of amino acids, carbohydrates, lipids, nucleic acids, proteins, steroids, and other biologically active molecules, In HPLC, mobile phase passes through columns under 10–400 atmospheric pressure, and with a high (0.1–5 cm//sec) flow rate. In this technique, use of small particles, and application of high pressure on the rate of solvent flow increases separation power, of HPLC and the analysis is completed within a short time.

Essential components of a HPLC device are solvent depot, high- pressure pump, commercially prepared column, detector, and recorder. Duration of separation is controlled with the aid of a computerized system, and material is accrued.

Applications of Chromatography

Chromatography is one of the important analytical techniques which play an important role in industries especially in pharmaceutical industries and chemical industry.

In Pharmaceutical sector, the main role is to identify and analyse samples for the presence of trace elements or chemical. Its major role is the separation of compounds based on their molecular weight and element composition. Chromatography also detects the unknown compounds and purity of mixture. Chromatography importance is growing its importance in drug development.

In Chemical industry, Chromatography is applied in testing water samples and also checks air quality. HPLC and GC are very much used for detecting various contaminants such as polychlorinated biphenyl (PCBs) in pesticides and oils.

Chromatography also used in various life sciences applications, such as:

- Food Industry: Chromatography has been playing a vital role in Food spoilage and additive detection. In addition it also helps in determining nutritional quality of food.

- Forensic Science: Gas Chromatography has been widely used in forensic pathology and crime scene testing like analysing blood and hair samples of crime place.
- Molecular Biology Studies: Various hyphenated techniques in chromatography such as EC-LC-MS are applied in study of metabolomics and proteomics along with nucleic acids research.

The other applications of chromatography especially HPLC is used in Protein Separation like Insulin Purification, Plasma Fractionation and Enzyme Purification and also in various departments like Fuel Industry, biotechnology and biochemical processes.

Gas Chromatography

Gas chromatography is a term used to describe the group of analytical separation techniques used to analyze volatile substances in the gas phase. In gas chromatography, the components of a sample are dissolved in a solvent and vaporized in order to separate the analytes by distributing the sample between two phases: a stationary phase and a mobile phase. The mobile phase is a chemically inert gas that serves to carry the molecules of the analyte through the heated column. Gas chromatography is one of the sole forms of chromatography that does not utilize the mobile phase for interacting with the analyte. The stationary phase is either a solid adsorbant, termed gas-solid chromatography (GSC), or a liquid on an inert support, termed gas-liquid chromatography (GLC).

In early 1900s, Gas chromatography (GC) was discovered by Mikhail Semenovich Tsvett as a separation technique to separate compounds. In organic chemistry, liquid-solid column chromatography is often used to separate organic compounds in solution. Among the various types of gas chromatography, gas-liquid chromatography is the method most commonly used to separate organic compounds. The combination of gas chromatography and mass spectrometry is an invaluable tool in the identification of molecules. A typical gas chromatograph consists of an injection port, a column, carrier gas flow control equipment, ovens and heaters for maintaining temperatures of the injection port and the column, an integrator chart recorder and a detector.

To separate the compounds in gas-liquid chromatography, a solution sample that contains organic compounds of interest is injected into the sample port where it will be vaporized. The vaporized samples that are injected are then carried by an inert gas, which is often used by helium or nitrogen. This inert gas goes through a glass column packed with silica that is coated with a liquid. Materials that are less soluble in the liquid will increase the result faster than the material with greater solubility.

In GLC, the liquid stationary phase is adsorbed onto a solid inert packing or immobilized on the capillary tubing walls. The column is considered packed if the glass or metal column tubing is packed with small spherical inert supports. The liquid phase adsorbs onto the surface of these beads in a thin layer. In a capillary column, the tubing walls are coated with the stationary phase or an adsorbant layer, which is capable of supporting the liquid phase. However, the method of GSC, has limited application in the laboratory and is rarely used due to severe peak tailing and the semi-permanent retention of polar compounds within the column. Therefore, the method of gas-liquid chromatography is simply shortened to gas chromatography and will be referred to as such here.

Instrumentation

Sample Injection

A sample port is necessary for introducing the sample at the head of the column. Modern injection techniques often employ the use of heated sample ports through which the sample can be injected and vaporized in a near simultaneous fashion. A calibrated microsyringe is used to deliver a sample volume in the range of a few microliters through a rubber septum and into the vaporization chamber. Most separations require only a small fraction of the initial sample volume and a sample splitter is used to direct excess sample to waste. Commercial gas chromatographs often allow for both split and splitless injections when alternating between packed columns and capillary columns. The vaporization chamber is typically heated 50 °C above the lowest boiling point of the sample and subsequently mixed with the carrier gas to transport the sample into the column.

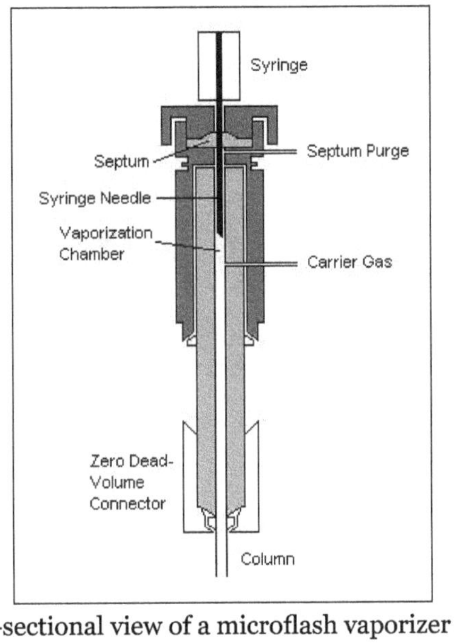

Figure: A cross-sectional view of a microflash vaporizer direct injector.

Carrier Gas

The carrier gas plays an important role, and varies in the GC used. Carrier gas must be dry, free of oxygen and chemically inert mobile-phase employed in gas chromatography. Helium is most commonly used because it is safer than, but comprable to hydrogen in efficiency, has a larger range of flow rates and is compatible with many detectors. Nitrogen, argon, and hydrogen are also used depending upon the desired performance and the detector being used. Both hydrogen and helium, which are commonly used on most traditional detectors such as Flame Ionization(FID), thermal conductivity (TCD) and Electron capture (ECD), provide a shorter analysis time and lower elution temperatures of the sample due to higher flow rates and low molecular weight. For instance, hydrogen or helium as the carrier gas gives the highest sensitivity with TCD because the difference in thermal conductivity between the organic vapor and hydrogen/helium is greater than other carrier gas. Other detectors such as mass spectroscopy, uses nitrogen or argon which has a much better advantage than hydrogen or helium due to their higher molecular weights, in which improve vacuum pump efficiency.

All carrier gasses are available in pressurized tanks and pressure regulators, gages and flow meters are used to meticulously control the flow rate of the gas. Most gas supplies used should fall between 99.995% - 99.9995% purity range and contain a low levels (< 0.5 ppm) of oxygen and total hydrocarbons in the tank. The carrier gas system contains a molecular sieve to remove water and other impurities. Traps are another option to keep the system pure and optimum sensitive and removal traces of water and other contaminants. A two stage pressure regulation is required to use to minimize the pressure surges and to monitor the flow rate of the gas. To monitor the flow rate of the gas a flow or pressure regulator was also require onto both tank and chromatograph gas inlet. This applies different gas type will use different type of regulator. The carrier gas is preheated and filtered with a molecular sieve to remove impurities and water prior to being introduced to the vaporization chamber. A carrier gas is typically required in GC system to flow through the injector and push the gaseous components of the sample onto the GC column, which leads to the detector.

Gas Recommendations for Capillary Columns

Detector	Carrier gas	Preferred makeup gas	Second choice	Detector, anode purge, or reference gas
Electron Capture	Hydrogen	Argon/Methane	Nitrogen	Anode purge must be same as makeup
	Helium	Argon/Methane	Nitrogen	
	Nitrogen	Nitrogen	Argon/Methane	
	Argon/Methane	Argon/Methane	Nitrogen	
Flame Ionization	Hydrogen	Nitrogen	Helium	Hydrogen and air for detector
	Helium	Nitrogen	Helium	
	Nitrogen	Nitrogen	Helium	
Flame Photometric	Hydrogen	Nitrogen		Hydrogen and air for detector
	Helium	Nitrogen		
	Nitrogen	Nitrogen		
	Argon	Nitrogen		
Nitrogen-Phosphorus	Helium	Nitrogen	Helium**	Hydrogen and air for detector
	Nitrogen	Nitrogen	Helium**	
Thermal Conductivity	Hydrogen*	Must be same as carrier and reference gas	Must be same as carrier and reference gas	Reference must be same as carrier and makeup
	Helium			
	Nitrogen			

* When using hydrogen with a thermal conductivity detector, vent the detector exhaust to a fume hood or a dedicated exhaust to avoid buildup of hydrogen gas.

**Helium is not recommended as a makeup gas at flow rates > 5 mL/min. Flow rates above 5 mL/min shorten detector life.

Figure: Gas Recommendations for Capillary Columns.

Gas Recommendations for Packed Columns

Detector	Carrier gas	Comments	Detector, anode purge, or reference gas
Electron Capture	Nitrogen	Maximum sensitivity	Nitrogen
	Argon/Methane	Maximum dynamic range	Argon/Methane
Flame Ionization	Nitrogen	Maximum sensitivity	Hydrogen and air for detector
	Helium	Acceptable alternative	
Flame Photometric	Hydrogen		Hydrogen and air for detector
	Helium		
	Nitrogen		
	Argon		
Nitrogen-Phosphorus	Helium	Optimum performance	Hydrogen and air for detector
	Nitrogen	Acceptable alternative	
Thermal Conductivity	Helium	General use	Reference must be same as carrier
	Hydrogen	Maximum sensitivity (Note A)	
	Nitrogen	Hydrogen detection (Note B)	
	Argon	Maximum hydrogen sensitivity (Note B)	

Note A: Slightly greater sensitivity than helium. Incompatible with some compounds.
Note B: For analysis of hydrogen or helium. Greatly reduces sensitivity for other compounds.

Figure: Gas Recommendations for Packed Columns.

Column Oven

The thermostatted oven serves to control the temperature of the column within a few tenths of a degree to conduct precise work. The oven can be operated in two manners: isothermal programming or temperature programming. In isothermal programming, the temperature of the column is held constant throughout the entire separation. The optimum column temperature for isothermal operation is about the middle point of the boiling range of the sample. However, isothermal programming works best only if the boiling point range of the sample is narrow. If a low isothermal column temperature is used with a wide boiling point range, the low boiling fractions are well resolved but the high boiling fractions are slow to elute with extensive band broadening. If the temperature is increased closer to the boiling points of the higher boiling components, the higher boiling components elute as sharp peaks but the lower boiling components elute so quickly there is no separation.

In the temperature programming method, the column temperature is either increased continuously or in steps as the separation progresses. This method is well suited to separating a mixture with a broad boiling point range. The analysis begins at a low temperature to resolve the low boiling components and increases during the separation to resolve the less volatile, high boiling components of the sample. Rates of 5-7 °C/minute are typical for temperature programming separations.

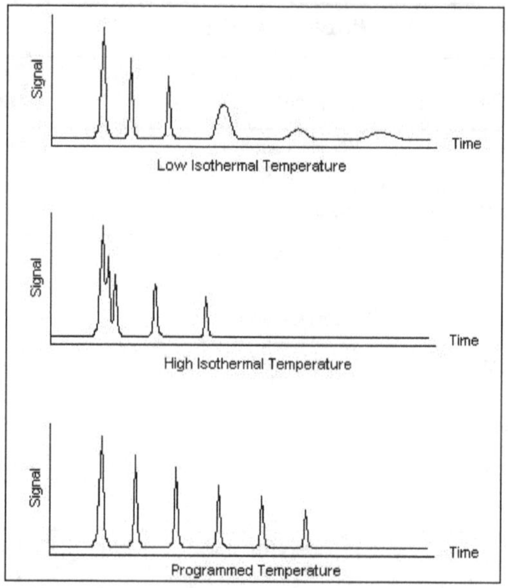

Figure: The effect of column temperature on the shape of the peaks.

Open Tubular Columns and Packed Columns

Open tubular columns, which are also known as capillary columns, come in two basic forms. The first is a wall-coated open tubular (WCOT) column and the second type is a support-coated open tubular (SCOT) column. WCOT columns are capillary tubes that have a thin layer of the stationary phase coated along the column walls. In SCOT columns, the column walls are first coated with a thin layer (about 30 micrometers thick) of adsorbant solid, such as diatomaceous earth, a material which consists of single-celled, sea-plant skeletons. The adsorbant solid is then treated with the liquid stationary phase. While SCOT columns are capable of holding a greater volume of stationary phase than a WCOT column due to its greater sample capacity, WCOT columns still have greater column efficiencies.

Most modern WCOT columns are made of glass, but T316 stainless steel, aluminum, copper and plastics have also been used. Each material has its own relative merits depending upon the application. Glass WCOT columns have the distinct advantage of chemical etching, which is usually achieved by gaseous or concentrated hydrochloric acid treatment. The etching process gives the glass a rough surface and allows the bonded stationary phase to adhere more tightly to the column surface.

One of the most popular types of capillary columns is a special WCOT column called the fused-silica wall-coated (FSWC) open tubular column. The walls of the fused-silica columns are drawn from purified silica containing minimal metal oxides. These columns are much thinner than glass columns, with diameters as small as 0.1 mm and lengths as long as 100 m. To protect the column, a polyimide coating is applied to the outside of the tubing and bent into coils to fit inside the thermostatted oven of the gas chromatography unit. The FSWC columns are commercially available and currently replacing older columns due to increased chemical inertness, greater column efficiency and smaller sampling size requirements. It is possible to achieve up to 400,000 theoretical plates with a 100 m WCOT column, yet the world record for the largest number of theoretical plates is over 2 million plates for 1.3 km section of column.

Packed columns are made of a glass or metal tubing which is densely packed with a solid support like diatomaceous earth. Due to the difficulty of packing the tubing uniformly, these types of columns have a larger diameter than open tubular columns and have a limited range of length. As a result, packed columns can only achieve about 50% of the efficiency of a comparable WCOT column. Furthermore, the diatomaceous earth packing is deactivated over time due to the semi-permanent adsorption of impurities within the column. In contrast, FSWC open tubular columns are manufactured to be virtually free of these adsorption problems.

	Type of Column			
	FSWC	WCOT	SCOT	Packed
Length	10 to 1000 m	10 to 1000 m	10 to 100 m	1 to 6 m
Inner Diameter	0.1 to 0.3 mm	0.25 to 0.75 mm	0.5 mm	2 to 4 mm
Efficiency (plates/m)	2000 to 4000	1000 to 4000	600 to 1200	500 to 1000
Sample Size	10 to 75 ng	10 to 1000 ng	10 to 1000 ng	10 to 10^6 ng
Pressure	Low	Low	Low	High
Speed	Fast	Fast	Fast	Slow
Inertness	Best	Good	Fair	Poor

Figure: Properties of gas chromatography columns.

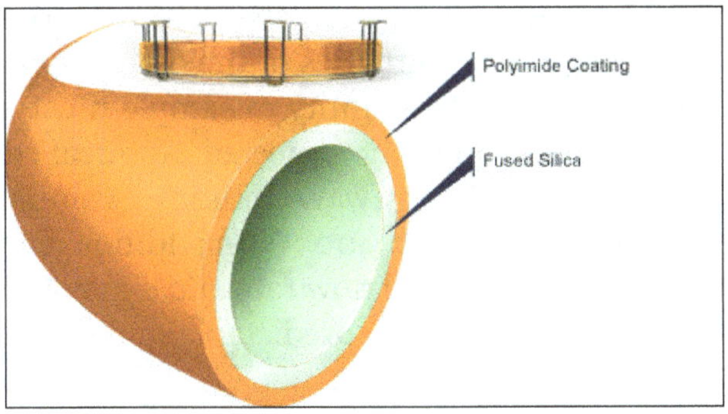

Figure: Computer Generated Image of a FSWC column (specialized for measuring BAC levels).

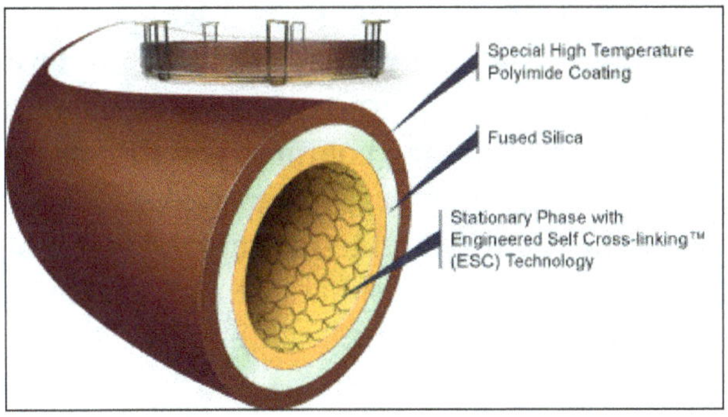

Figure: Computer Generated Image of a FSWC column (specialized to withstand extreme heat).

Different types of columns can be applied for different fields. Depending on the type of sample, some GC columns are better than the others. For example, the FSWC column shown in Figure 5 is designed especially for blood alcohol analysis. It produces fast run times with baseline resolution of key components in less than 3 minutes. Moreover, it displays enhanced resolutions of ethanol and acetone peaks, which helps with determining the BAC levels. This particular column is known as Zebron-BAC and it made with polyimide coating on the outside and the inner layer is made of fused silica and the inner diameter ranges from .18 mm to .25 mm. There are also many other Zebron brand columns designed for other purposes.

Another example of a Zebron GC column is known as the Zebron-inferno. Its outer layer is coated with a special type of polyimide that is designed to withstand high temperatures. As shown in figure, it contains an extra layer inside. It can withstand up to 430 °C to be exact and it is designed to provide true boiling point separation of hydrocarbons distillation methods. Moreover, it is also used for acidic and basic samples.

Detection Systems

The detector is the device located at the end of the column which provides a quantitative measurement of the components of the mixture as they elute in combination with the carrier gas. In theory, any property of the gaseous mixture that is different from the carrier gas can be used as a detection method. These detection properties fall into two categories: bulk properties and specific properties. Bulk properties, which are also known as general properties, are properties that both the carrier gas and analyte possess but to different degrees. Specific properties, such as detectors that measure nitrogen-phosphorous content, have limited applications but compensate for this by their increased sensitivity.

Each detector has two main parts that when used together they serve as transducers to convert the detected property changes into an electrical signal that is recorded as a chromatogram. The first part of the detector is the sensor which is placed as close the the column exit as possible in order to optimize detection. The second is the electronic equipment used to digitize the analog signal so that a computer may analyze the acquired chromatogram. The sooner the analog signal is converted into a digital signal, the greater the signal-to-noise ratio becomes, as analog signal are easily susceptible to many types of interferences.

An ideal GC detector is distinguished by several characteristics. The first requirement is adequate sensitivity to provide a high resolution signal for all components in the mixture. This is clearly an idealized statement as such a sample would approach zero volume and the detector would need infinite sensitivity to detect it. In modern instruments, the sensitivities of the detectors are in the range of 10^{-8} to 10^{-15} g of solute per second. Furthermore, the quantity of sample must be reproducible and many columns will distort peaks if enough sample is not injected. An ideal column will also be chemically inert and should not alter the sample in any way. Optimized columns will be able to withstand temperatures in the range of -200 °C to at least 400 °C. In addition, such a column would have a short linear response time that is independent of flow rate and extends for several orders of magnitude. Moreover, the detector should be reliable, predictable and easy to operate.

Understandably, it is not possible for a detector meet all of these requirements. The next subsections will discuss some of the more common types of gas chromatography detectors and the relative advantages and/or disadvantages of each.

Type of Detector	Applicable Samples	Detection Limit
Mass Spectrometer (MS)	Tunable for any sample	.25 to 100 pg
Flame Ionization (FID)	Hydrocarbons	1 pg/s
Thermal Conductivity (TCD)	Universal	500 pg/ml
Electron-Capture (ECD)	Halogenated hydrocarbons	5 fg/s
Atomic Emission (AED)	Element-selective	1 pg
Chemiluminescence (CS)	Oxidizing reagent	Dark current of PMT
Photoionization (PID)	Vapor and gaseous Compounds	.002 to .02 µg/L

Table: Typical gas chromatography detectors and their detection limits.

Mass Spectrometry Detectors

Mass Spectrometer (MS) detectors are most powerful of all gas chromatography detectors. In a GC/MS system, the mass spectrometer scans the masses continuously throughout the separation. When the sample exits the chromatography column, it is passed through a transfer line into the inlet of the mass spectrometer. The sample is then ionized and fragmented, typically by an electron-impact ion source. During this process, the sample is bombarded by energetic electrons which ionize the molecule by causing them to lose an electron due to electrostatic repulsion. Further bombardment causes the ions to fragment. The ions are then passed into a mass analyzer where the ions are sorted according to their m/z value, or mass-to-charge ratio. Most ions are only singly charged.

The Chromatogram will point out the retention times and the mass spectrometer will use the peaks to determine what kinds of molecules are exist in the mixture. The figure below represents a typical mass spectrum of water with the absorption peaks at the appropriate m/z ratios.

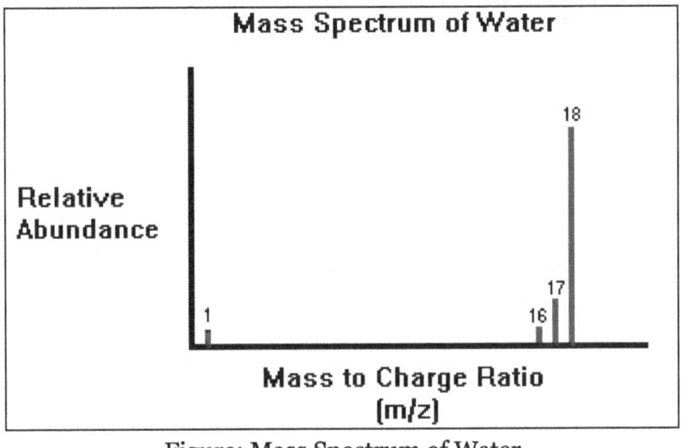

Figure: Mass Spectrum of Water.

Instrumentation

One of the most common types of mass analyzer in GC/MS is the quadrupole ion-trap analyzer, which allows gaseous anions or cations to be held for long periods of time by electric and magnetic fields. A simple quadrupole ion-trap consists of a hollow ring electrode with two grounded end-cap electrodes as seen in figure. Ions are allowed into the cavity through a grid in the upper end cap. A variable radio-frequency is applied to the ring electrode and ions with an appropriate m/z value orbit around the cavity. As the radio-frequency is increased linearly, ions of a stable m/z value are ejected by mass-selective ejection in order of mass. Ions that are too heavy or too light are destabilized and their charge is neutralized upon collision with the ring electrode wall. Emitted ions then strike an electron multiplier which converts the detected ions into an electrical signal. This electrical signal is then picked up by the computer through various programs. As an end result, a chromatogram is produced representing the m/z ratio versus the abundance of the sample.

GC/MS units are advantageous because they allow for the immediate determination of the mass of the analyte and can be used to identify the components of incomplete separations. They are rugged, easy to use and can analyze the sample almost as quickly as it is eluted. The disadvantages of mass spectrometry detectors are the tendency for samples to thermally degrade before detection and the end result of obliterating all the sample by fragmentation.

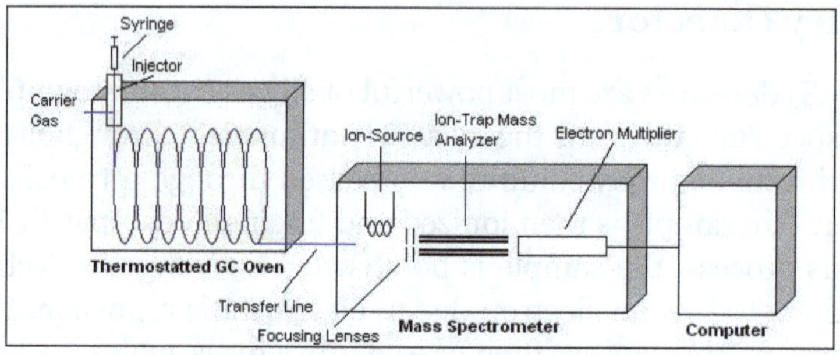

Figure: Schematic of the GC/MS system.

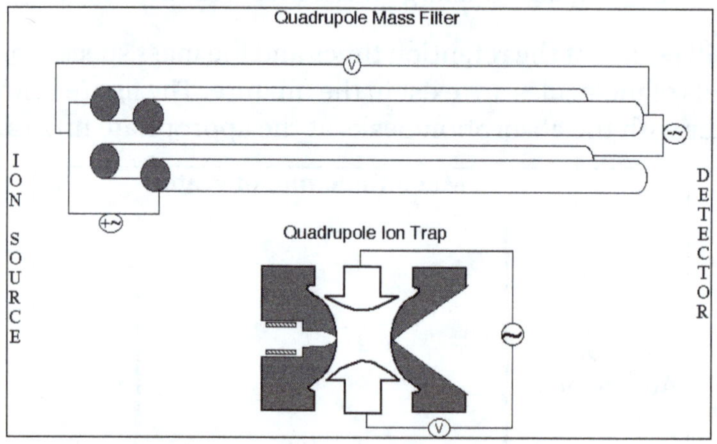

Figure: Arrangement of the poles in Quadrupole and Ion Trap Mass spectrometers.

Flame Ionization Detectors

Flame ionization detectors (FID) are the most generally applicable and most widely used detectors. In a FID, the sample is directed at an air-hydrogen flame after exiting the column. At the high temperature of the air-hydrogen flame, the sample undergoes pyrolysis, or chemical decomposition through intense heating. Pyrolized hydrocarbons release ions and electrons that carry current. A high-impedance picoammeter measures this current to monitor the sample's elution.

It is advantageous to used FID because the detector is unaffected by flow rate, noncombustible gases and water. These properties allow FID high sensitivity and low noise. The unit is both reliable and relatively easy to use. However, this technique does require flammable gas and also destroys the sample.

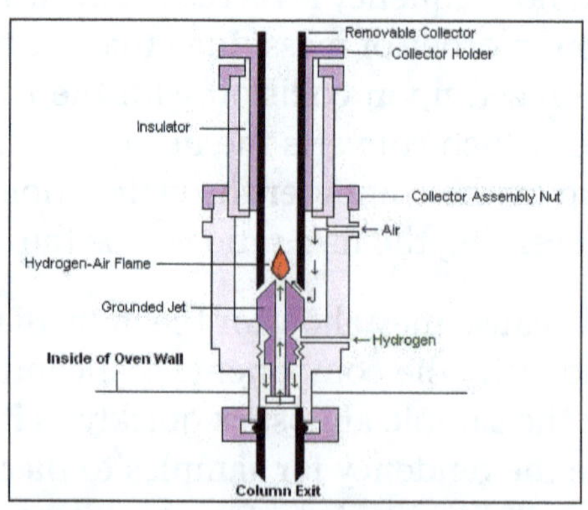

Figure: Schematic of a typical flame ionization detector.

Thermal Conductivity Detectors

Thermal conductivity detectors (TCD) were one the earliest detectors developed for use with gas chromatography. The TCD works by measuring the change in carrier gas thermal conductivity caused by the presence of the sample, which has a different thermal conductivity from that of the carrier gas. Their design is relatively simple, and consists of an electrically heated source that is maintained at constant power. The temperature of the source depends upon the thermal conductivities of the surrounding gases. The source is usually a thin wire made of platinum, gold. The resistance within the wire depends upon temperature, which is dependent upon the thermal conductivity of the gas.

TCDs usually employ two detectors, one of which is used as the reference for the carrier gas and the other which monitors the thermal conductivity of the carrier gas and sample mixture. Carrier gases such as helium and hydrogen has very high thermal conductivities so the addition of even a small amount of sample is readily detected.

The advantages of TCDs are the ease and simplicity of use, the devices' broad application to inorganic and organic compounds, and the ability of the analyte to be collected after separation and detection. The greatest drawback of the TCD is the low sensitivity of the instrument in relation to other detection methods, in addition to flow rate and concentration dependency.

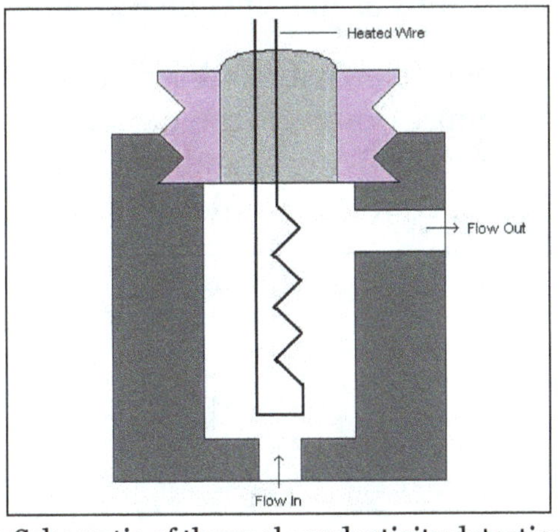

Figure:Schematic of thermal conductivity detection cell.

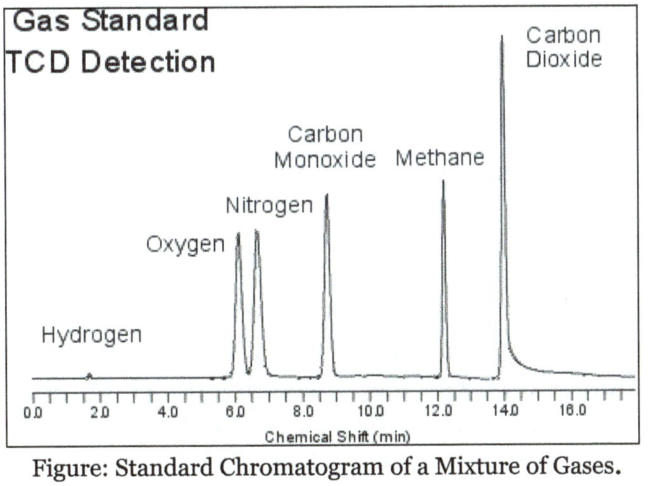

Figure: Standard Chromatogram of a Mixture of Gases.

Chromatogram

Figure represents a standard chromatogram produced by a TCD detector. In a standard chromatogram regardless of the type detector, the x-axis is the time and the y-axis is the abundance or the absorbance. From these chromatograms, retention times and the peak heights are determined and used to further investigate the chemical properties or the abundance of the samples.

Electron-capture Detectors

Electron-capture detectors (ECD) are highly selective detectors commonly used for detecting environmental samples as the device selectively detects organic compounds with moieties such as halogens, peroxides, quinones and nitro groups and gives little to no response for all other compounds. Therefore, this method is best suited in applications where traces quantities of chemicals such as pesticides are to be detected and other chromatographic methods are unfeasible.

The simplest form of ECD involves gaseous electrons from a radioactive emitter in an electric field. As the analyte leaves the GC column, it is passed over this emitter, which typically consists of nickle-63 or tritium. The electrons from the emitter ionize the nitrogen carrier gas and cause it to release a burst of electrons. In the absence of organic compounds, a constant standing current is maintained between two electrodes. With the addition of organic compounds with electronegative functional groups, the current decreases significantly as the functional groups capture the electrons.

The advantages of ECDs are the high selectivity and sensitivity towards certain organic species with electronegative functional groups. However, the detector has a limited signal range and is potentially dangerous owing to its radioactivity. In addition, the signal-to-noise ratio is limited by radioactive decay and the presence of O_2 within the detector.

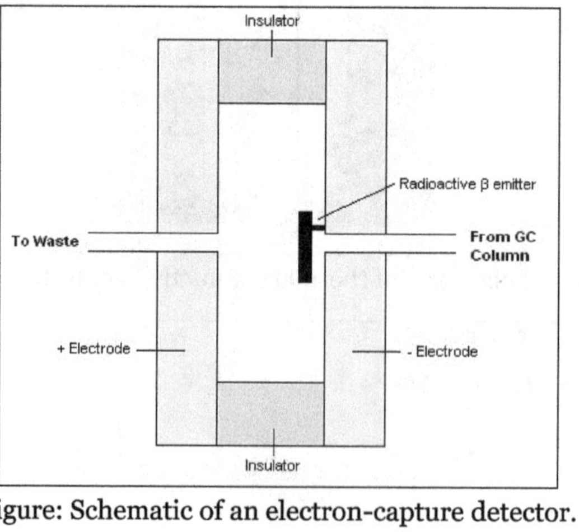

Figure: Schematic of an electron-capture detector.

Atomic Emission Detectors

Atomic emission detectors (AED), one of the newest addition to the gas chromatographer's arsenal, are element-selective detectors that utilize plasma, which is a partially ionized gas, to atomize all of the elements of a sample and excite their characteristic atomic emission spectra. AED is an extremely powerful alternative that has a wider applicability due to its based on the detection of atomic emissions. There are three ways of generating plasma: microwave-induced plasma (MIP), inductively coupled plasma (ICP) or direct current plasma (DCP). MIP is the most commonly employed form and is used with a positionable diode array to simultaneously monitor the atomic emission spectra of several elements.

Instrumentation

The components of the Atomic emission detectors include 1) an interface for the incoming capillary GC column to induce plasma chamber,2) a microwave chamber, 3) a cooling system, 4) a diffration grating that associated optics, and 5) a position adjustable photodiode array interfaced to a computer.

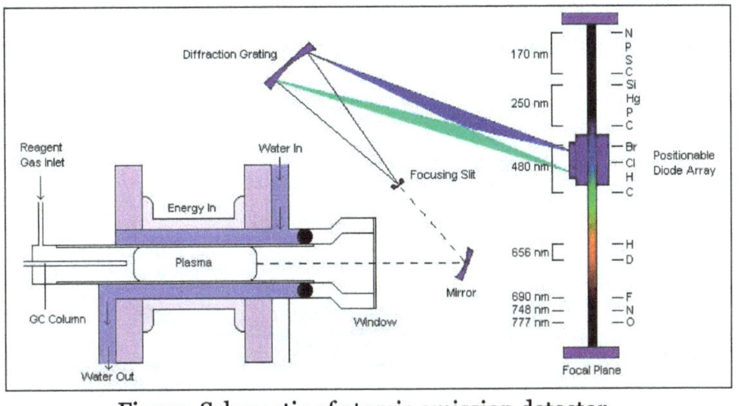

Figure: Schematic of atomic emission detector.

GC Chemiluminescence Detectors

Chemiluminescence spectroscopy (CS) is a process in which both qualitative and quantitative properties can be determined using the optical emission from excited chemical species. It is very similar to AES, but the difference is that it utilizes the light emitted from the energized molecules rather than just excited molecules. Moreover, chemiluminescence can occur in either the solution or gas phase whereas AES is designed for gaseous phases. The light source for chemiluminescence comes from the reactions of the chemicals such that it produces light energy as a product. This light band is used instead of a separate source of light such as a light beam.

Like other methods, CS also has its limitations and the major limitation to the detection limits of CS concerns with the use of a photomultiplier tube (PMT). A PMT requires a dark current in it to detect the light emitted from the analyte.

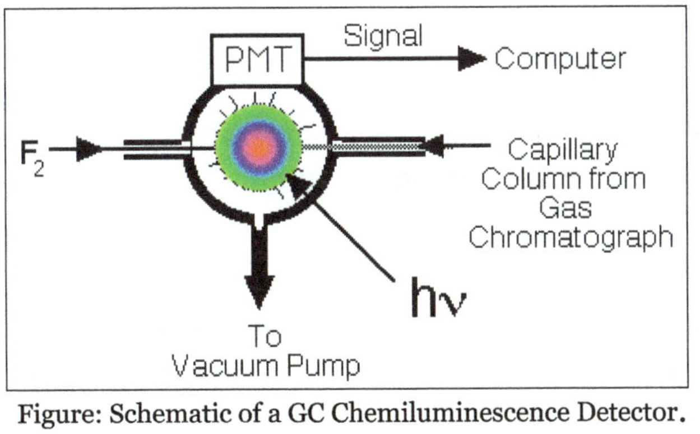

Figure: Schematic of a GC Chemiluminescence Detector.

Photoionization Detectors

Another different kind of detector for GC is the photoionization detector which utilizes the properties of chemiluminescence spectroscopy. Photoionization detector (PID) is a portable vapor and gas detector that has selective determination of aromatic hydrocarbons, organo-heteroatom, inorganice species and other organic compounds. PID comprise of an ultrviolet lamp to emit photons that are absorbed by the compounds in an ionization chamber exiting from a GC column. Small fraction of the analyte molecules are actually ionized, nondestructive, allowing confirmation analytical results through other detectors. In addition, PIDs are available in portable hand-held models and in a number of lamp configurations. Results are almost immediate. PID is used commonly to detect VOCs in soil, sediment, air and water, which is often used to detect contaminants in ambient air and soil. The disavantage of PID is unable to detect certain hydrocarbon that has low molecular weight, such as methane and ethane.

Instrumentation

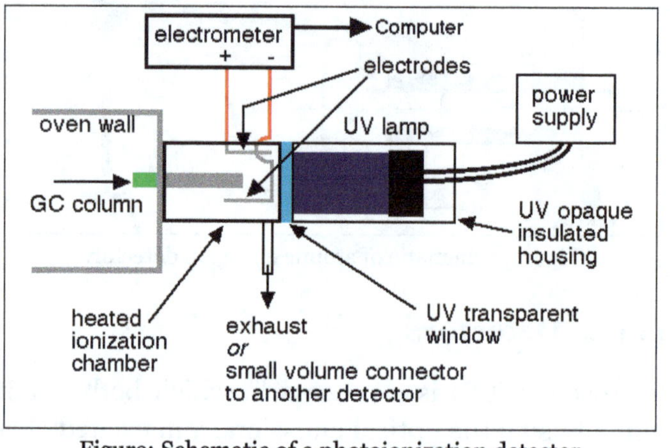

Figure: Schematic of a photoionization detector.

Limitations:

1. Not suitable for detecting semi-volatile compounds.
2. Only indicates if volatile organic compounds are presents.
3. High concentration so methane is required for higher performance.
4. Frequent calibration is required.
5. Units of parts per million range.
6. Environmental distraction, especially water vapor.
7. Strong electrical field's Rapid variation in temperature at the detector and naturally occurring compounds may affect instrumental signal.

Applications

Gas chromatography is a physical separation method in where volatile mixtures are separated. It can be used in many different fields such as pharmaceuticals, cosmetics and even environmental toxins. Since the samples have to be volatile, human breathe, blood, saliva and other secretions containing large amounts of organic volatiles can be easily analyzed using GC. Knowing the amount of which compound is in a given sample gives a huge advantage in studying the effects of human health and of the environment as well.

Air samples can be analyzed using GC. Most of the time, air quality control units use GC coupled with FID in order to determine the components of a given air sample. Although other detectors are useful as well, FID is the most appropriate because of its sensitivity and resolution and also because it can detect very small molecules as well.

GC/MS is also another useful method which can determine the components of a given mixture using the retention times and the abundance of the samples. This method be applied to many pharmaceutical applications such as identifying the amount of chemicals in drugs. Moreover, cosmetic manufacturers also use this method to effectively measure how much of each chemical is used for their products.

Liquid Chromatography

Liquid chromatography (LC) is an analytical chromatographic technique that is useful for separating ions or molecules that are dissolved in a solvent.

If the sample solution is in contact with a second solid or liquid phase, the different solutes will interact with the other phase to differing degrees due to differences in adsorption, ion-exchange, partitioning, or size.

These differences allow the mixture components to be separated from each other by using these differences to determine the transit time of the solutes through a column.

Instrumentation Simple liquid chromatography consists of a column with a fritted bottom that holds a stationary phase in equilibrium with a solvent. Typical stationary phases (and their interactions with the solutes) are: solids (adsorption), ionic groups on a resin (ion-exchange), liquids on an inert solid support (partitioning), and porous inert particles (size-exclusion).

The mixture to be separated is loaded onto the top of the column followed by more solvent. The different components in the sample mixture pass through the column at different rates due to differences in their partitioning behavior between the mobile liquid phase and the stationary phase. The compounds are separated by collecting aliquots of the column effluent as a function of time.

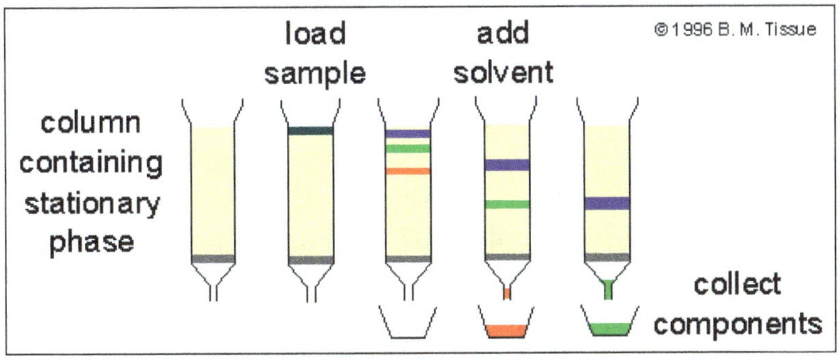

Schematic of a simple liquid chromatographic separation.

Conventional LC is most commonly used in preparative scale work to purify and isolate some components of a mixture. It is also used in ultra-trace separations where small disposable columns are used once and then discarded. Analytical separations of solutions for detection or quantification typically use more sophisticated high-performance liquid chromatography instruments.

Extraction

Solvent extraction technique is a part of analytical chemistry and has been recognized as an excellent separation method because of its ease, simplicity, speed, and wide scope. Utilizing apparatus no more complicated than a separately funnel, requiring just several minutes, at the most to perform, applicable both to trace and macro levels of metals, extraction procedures offers much to the analytical chemist. A further advantage of the extraction method over the widely used precipitation method lies in the cleaner separations that can be achieved by the former. With the later method contamination of precipitates by co-precipitation phenomena is a decided limitation which is minimized only with difficulty, whereas the analog of co-precipitation, i.e., co-extraction, is almost unknown in solvent extraction.

After world war second, chemists were engaged in atomic energy research, encountered the problem of separating and purifying almost all the elements, in amounts ranging from very low traces up to the usual micro levels. Among the solution to these difficult problems were the precipitation method and the ion exchange method. The former was found to be suitable for the treatment of materials in solution at high electrolyte concentrations and the later for the treatment of the ions in dilute solution or for the separation of chemically similar elements such as lanthanides and transplutonium elements.

However, the solvent extraction method proved to be most effective and attractive to separate these metals, because in this method, the separation is almost complete than precipitation method and operation time is very less as compared to ion exchange method. With proper choice of extracting agents, this technique can achieve group separation or selective separation of trace elements with high efficiencies. In analytical applications, solvent extraction may serve the following three purposes:

i) Pre-concentration of trace elements.

ii) Elimination of matrix interference.

iii) Differentiation of chemical species.

Solvent extraction or liquid-liquid extraction by high molecular weight organic amines has become increasingly popular in recent years in studying metal complexes. Extraction by these organic amines combine many of the advantages of both solvent extraction and ion exchange. The main interest of metal extraction by the high molecular weight amines lies in their selectivity towards anionic metal complexes, reversibly formed in an aqueous solution. Hence they are generally referred to as "liquid anion exchangers". The extent of extraction by the organic bases depends on their nature, structure, size, concentration and the nature of the organic solvent used as diluent.

1) The solubility of primary long chain amines in non-polar solvents increases with increasing chain length, secondary amines are generally highly soluble in non-polar solvents and

sparingly soluble in highly polar solvents. Tertiary amines are completely miscible with non-polar solvents at room temperature and sparingly soluble in polar solvents.

2) Extractive power of the alkylamines generally increases from primary to secondary to tertiary to quaternary amines. The trend may be exceptionally changed if there is extraction by bulky amines, where steric factor play an important role.

3) Third phase formation:

   a) The third phase formation, a second organic one, is more common in systems where the diluent is an aliphatic hydrocarbon. Aromatic and some derivatives of aliphatic hydrocarbons show the phenomenon usually at high organic phase loading.

   b) The splitting of organic phase is less common, when straight chain alkylamines are used.

   c) Formation of third phase is most common, characteristic of amine sulphate systems. The compatibility increases in the order: sulphate >bisulphate > chloride > nitrate.

   d) The formation of the third phase is temperature dependent. Literature in solvent extraction is reviewed by Freiser every two years in 'Analytical Chemistry'. Morrison and Freiser wrote a comprehensive monograph on solvent extraction in analytical chemistry. The solvent extraction of metal chelate complexes was reviewed by Starry and Zoltov. Marcus and Kertes and De et al, the applications were compiled by Sakine and Hasegawa. The extraction of many metal ions from various aqueous solutions by high molecular weight amines have been reviewed by Khopkar and Green.

High molecular weight amine systems used for liquid-liquid extraction of many metals from different aqueous solutions are listed as:

Adogen 364, Alamine 336 (tricarpyl amine), Aliquat 336 (tricarpyl methyl ammonium chloride), Amerlite LA-1 (N-dodecyl trialkyl methyl amine), LA-2 (N-lauryl trialkyl methyl amine), n-octyl aniline, TBA (tribenzyl amine), TOA (tri-n-octyl amine), TIOA (tri-iso-octyl amine), TLA (tri-n-lauryl amine).

Basic Principles of Solvent Extraction Method

An extractant, is a substance primarily responsible for the transfer of a solute (here metal) from one phase to the other. The extractant is dissolved in a suitable diluent and together act as a solvent. The diluent is immiscible with other phase which is usually water. The extractant reacts with the solute by solvation/chelation/ion pair formation etc. to extract from the aqueous phase. The distribution equilibrium between two phases is governed by Gibbs phase rule, given by,

$$P + V = C + 2$$

Where,

P = is the number of phases,

V = is the variance or degree of freedom,

C = is the number of components.

In solvent extraction, we have P=2 two phases namely aqueous and organic phase, the component C=1, viz. solute, in solvent and water phase and at constant temperature and pressure P=1, thus, we therefore have,

$$2+1=1+2 \qquad i.e. \qquad P+V=C+2$$

According to Nernst distribution law,

If $[X]_1$ is concentration of solute in phase 1 and $[X]_2$ is the concentration of solute in phase 2 at equilibrium:

$$K_D = \frac{[X]1}{[X]2}$$

Where K_D is called as the partition coefficient, this partition or distribution coefficient is independent of the total solute concentration in either of the phases. In the above expression for K_D, we have not considered the activity coefficient of the species in the organic as well as in the aqueous phase. We, therefore, use the term distribution ratio (D) to account for the total concentration of species in the two phases.

Distribution Ratio (D)

The distribution of a solute between two immiscible solvents in contact to each other can be described by the distribution ratio "D".

$$D = \frac{[X]_1}{[X]_2}$$

Where $[X]$ represents the stochiometric or formal concentration of a substance X and the subscripts 1 and 2 refer to the two phases. Since in most cases, two-phase system is of analytical interest, an organic solvent and aqueous are involved, D will be understood to be;

$$D = \frac{[X]_{org}}{[X]_{aq}}$$

The subscript org and aq refer to the organic and aqueous phases respectively. Distribution ratio 'D' is dimensionless quantity, separation of two solutes by solvent extraction is expressed by the term, separation factor (α), which is related to individual distribution ratios,

$$\alpha = \frac{D_A}{D_B}$$

D_A and D_B are the respective distribution ratios of solute A and B.

In the simplest extraction case, the distribution ratio is constant in accordance with the classical Nernst distribution law, a solute will distribute itself between two essentially immiscible solvents so that at equilibrium, the ratio of the concentrations of the solute in the two phases at a particular temperature will be constant, provided the solute is not involved in chemical interactions in either phases. For such a solute, then $K_D = D$. Now in practical work, instead of using term K_D or D one prefers to use term percentage extraction (E).

Percent Extraction (%E)

The more commonly used term for expressing the extraction efficiency by analytical chemist is the percent extraction "E", which is related to "D" as,

$$\%\text{Extraction}(E) = \frac{100D}{D + V_{aq} / V_{org}}$$

Where, V represent solvent volume and the other quantities remain as previously defined. The percent extraction may be seen to vary with the volume ratio of the two phases as well as with D.

Classification of Extraction Systems

The process of metal extraction is based on the formation of neutral metal chelate. All types of chelating agents find useful applications in metal extraction procedures. Various extraction systems can be classified in several ways. The classical one is based on the nature of the extracted species. The present day classification is based upon the process of extraction. Thus, based upon the process of extraction, extraction systems can be classified into four major classes viz.,

a) Chelate extraction.

b) Extraction by solvation.

c) Extraction involving ion pair formation.

d) Synergic extraction.

All the above extractions are based on the fact that neutral or uncharged species are extracted easily in organic solvents. These are described here briefly:

Chelate Extraction

In this class, extraction proceeds by the process of formation of chelate or closed ring structure between the chelating agent and the metal ion to be extracted.

e.g. i) The extraction of Uranium with 8-hydroxyquinoline in chloroform.

ii) The extraction of Iron with cupferron in carbon tetrachloride.

Extraction by Solvation

In this class, the extraction proceeds by the process of solvation of the species which is extracted into organic phase. Oxygenated organic solvents such as alcohols (C-OH), ketones, ethers and esters show some basicity because of the lone pair of electron on the oxygen atom and can therefore directly solvate protons and metal ions and bring about their extraction.

E.g. i) The extraction of Uranium with tributyl phosphate from nitric acid.

ii) The extraction of Iron(III) with diethyl ether from hydrochloric acid.

Extraction involving ion pair formation

The extraction proceeds with the formation of neutral uncharged species which in turn gets extracted in to the organic phase. The best example of this is the extraction of Scandium and Uranium with trioctyl amine from mineral acids. In this case an ion pair is formed between complex of metal ion with high molecular weight amine and anionic species of mineral acids.

Synergic Extraction

In this case, there is enhancement in the extraction on account of use of two extractants.

e.g. the extraction of Uranium with tributylphosphate (TBP) as well as 2- thionyltrifluroacetone (TTA).

Methods of Extraction

Three basic methods of liquid-liquid extraction are generally utilized in the analytical laboratory.

1. Batch Extraction

Batch extraction, the simplest and most commonly used method, consists of extracting the solute from one immiscible layer in to other by shaking the two layers until equilibrium is attained, after which the layers are allowed to settle before sampling. This is commonly used on the small scale in chemical laboratories. The most commonly employed apparatus for performing a batch extraction is a separatory funnel. The batch extractions may also be used with advantage when the distribution ratio is large.

2. Continuous Extraction

The second type, continuous extraction, makes use of a continuous flow of immiscible solvent through the solution or a continuous countercurrent flow of both phases. Continuous extractions are particularly applicable when the distribution ratio is relatively small. Continuous extraction device operate on the same general principle, which consist of distilling the extracting solvent from a boiler flask and condencing it and passing it continuously through the solution being extracted. The extracting liquid separates out and flows back into the receiving flask, where it is again evaporated and recycled while the extracted solute remains in the receiving flask. When the solvent cannot easily be distilled, a continuous supply of fresh solvent may be added from a reservoir.

3. Countercurrent Extractions

Extraction by continuous countercurrent distribution is the third general type and is used primarily for fractionation purposes. The separation through continuous countercurrent method is achieved by virtue of the density difference between the fluids in contact. In vertical columns, the denser phase enters at the top and flows downwards while the less dense phase enters from the bottom and flows upwards. The choice of method to be employed will depend primarily upon the value of the distribution ratio of the solute of interest, as well as on the separation factors of the interfering materials.

Factors Influencing the Extraction Efficiency

Primary requirement of solvent extraction for separation /removal purposes is a high distribution ratio of the solute of interest between the two liquid phases. It is useful to employ a number of different techniques for enhancing the distribution ratio. It depends on the nature of the species being extracted and extraction system. The attainment of selectivity in an extraction procedure is also very important. Some of the factors, which affect the distribution of solute of interest, are given below.

A. Choice of solvent

The most important consideration in the selection of a solvent for use in a particular extraction procedure is the extractability of the element of interest. For subsequent analytical processing a consideration of the solubility of the solute in particular solvent, the ease of recovery of the solvent or the ease of recovery of the solute from the solvent is very important. Thus, the boiling point of the solvent or the ease of stripping by chemical reagents enters into selection of a solvent when the possibility of a choice existed. Similarly, the degree of miscibility of the two phases, the relative specific gravities, viscosities and tendency to form emulsion should be considered. From the point of view of safety, the toxicity and the flammability of the solvent must be considered.

Use of a suitable solvent for effective separation is very important. Metal chelates and many organic molecules, being essentially covalent compounds do not impose many restrictions on the solvent and the general rules of solubility are the great use. In ion association systems and particularly in oxonium type ions, the role of solvents is very important. This is due to involvement of solvent in the formation of extractable species.

B. Acidity of an aqueous phase

The extractability of metal complexes is greatly influenced by the acidity of an aqueous phase, so it is necessary to assure optimum concentration of H+ ions for maximum extraction. In the case of chelate extraction, the chelating reagent concentration is maintained constant; the distribution of the metal in a system is a function of pH. For this reason, curves of extractability versus pH at constant reagent concentration are of great analytical significance. Sometimes it is possible to achieve the desired characteristics of a solvent by employing a mixed solvent system.

C. Stripping

Stripping is the removal of the extracted solute from the organic phase for further processing or analysis. In many colorimetric procedures and even radioactive techniques, the concentration of solute is determined directly in the organic phase. However, where further separation steps are required, it is necessary to remove the solute from the organic layer to more stable medium. When organic layer is on the steam bath, care should be taken, to avoid loss of volatile solute during evaporation. Addition of acid to water before evaporation helps to break the chelate complexes, thereby causing the metal ion to enter the aqueous phase. In the process of destroying the residual organic matter, hydrochloric acid, nitric acid, perchloric acid or aqua regia is used.

The usual procedure is to shake the organic layer with a volume of water alone or water containing an appropriate concentration of acid, an oxidizing or reducing agent or masking agent. The metal ion is then back extracted in the stripping aqueous phase.

The conditions employed depends upon metal ion and the particular extraction system and are such that they promote the reversal of extraction. Pure water or water adjusted to an appropriate pH/ molarity of acid are the more popular and convenient stripping agents. Washing the organic layer with an oxidizing and reducing agent changes the metal ion to be stripped in an oxidation state in which it is not extracted under the specific conditions.

D. Use of Masking Agents

In the extraction procedures for metal pairs that are difficult to separate; masking or sequestering agents are introduced to improve the separation factor. Masking agents are themselves metal-complexing agents, which serve to prevent a particular metal from taking part in their usual reaction, and thus to remove their interference without the necessity of actual separation. In solvent extraction, masking agents are used to prevent certain metals from forming extractable complexes and increasing the selectivity of the method. In extraction of metals the application of masking agents, which include cyanide, tartarate, citrate, fluoride, and EDTA. The selection of a particular masking agent largely depends on the acidity of the system and the complex formation constants of the metal with both the masking and the extraction agent.

EDTA, which has been proved as most useful masking agent, forms anionic complexes with a number of metal ions.

E. Salting-out Agents

The term salting-out agent is applied to those electrolytes whose addition greatly enhances the extractability of complexes. The function of salting-out agent would be primarily of providing a higher concentration of complex and thus improve the extraction. Water is probably bound as a shell of oriented water dipoles around the ion and thus becoming unavailable as "free solvent".

Addition of salting-out agents decreases the dielectric constant of the aqueous phase, which favors the formation of the ion association complexes. Salting-out agents have been used with great success in separation involving the halide and thiocynate systems.

In addition to enhancement of the extraction of the metal of interest using saltingout agents, it is also possible to decrease the extraction of impurities in the system. Thus, it is necessary to choose an agent that produces a favorable separation factor between the element of interest and the impurities. However, it must be remembered that anomalies sometimes result from specific interaction effects. Aluminium or calcium salts are strong salting-out agents, whereas ammonium salts are much weaker but analytically more convenient.

F. Backwashing

Backwashing is an auxiliary technique used with batch extractions to influence quantitative separations of elements. The combined organic phases from several extractions of the original aqueous phase contain practically all the desired elements and possibly some of the impurities that have been extracted to a much smaller extent. This combined organic phase when shaken with one or more small portions of a fresh aqueous phase containing the optimum reagent/salting agent concentration, acidity, etc., will result in a redistribution of the impurities in favor of the aqueous phase since their distribution ratios are low. Under optimum conditions, most of the elements of interest will remain in the organic layer, since their distribution ratios are high.

This technique is analogous in many respects to the re-precipitation step in a gravimetric precipitation procedure. With the proper conditions, most of the impurities can be removed by this backwashing operation, with neglisible loss of the main component, thereby attaining a selective operation.

G. Variation of Oxidation State

The selectivity of an extraction is increased by the modification of oxidation states of the interfering ions present in solution, in order to prevent the formation of their extractable metal complexes e.g. reduction of Cerium(IV) to Cerium(III) prevents extraction of this element from nitrate media, the extraction of Iron(III) from chloride solutions can be prevented by reduction to Iron(II), which is not extractable. Similarly, Antimony(V) may be reduced to the tetravalent state to suppress its extraction. Conversely, it is important in the preparation of a solution for extraction to adjust the proper valence state of metal ion required for the formation of the complex in order to ensure complete extraction of that element. Selectivity can also be achieved by variation of the oxidation state of the co-extracted interfering ions during the stripping operation.

H. Synergic Extraction

Synergism is defined as the combined action of two complexing reagents, which is greater than the sum of the actions of the individual reagents used alone. An example of the synergic extraction of Ce(III) with picrolonic acid and benzo-15-crown-5.

I. Use of Organic Acid Media

Organic acid media are having ability of controlling the concentration of the complexing ligand, is one of the unique application, the ease of adjustment of pH and the wide difference in pH at which various metal ions form anionic complexes. The comparative ease of stripping of the complexes from the organic phase can be achieved by fully exploiting the differences in reactivity of various metals to backwash in the aqueous phase by mineral acid. It is known that organic acid media offers better separation of metals possibly due to high stability of metal organic acids complexes.

Solid-Phase Extraction

SPE is a sample preparation technology that uses solid particle, chromatographic packing material, usually contained in a cartridge type device, to chemically separate the different components of a sample. Samples are nearly always in the liquid state [although specialty applications may be run with some samples in the gas phase]. Figure shows a sample, which appears black, being processed on a SPE device so that the individual dye compounds, which make up the sample, are chromatographically separated.

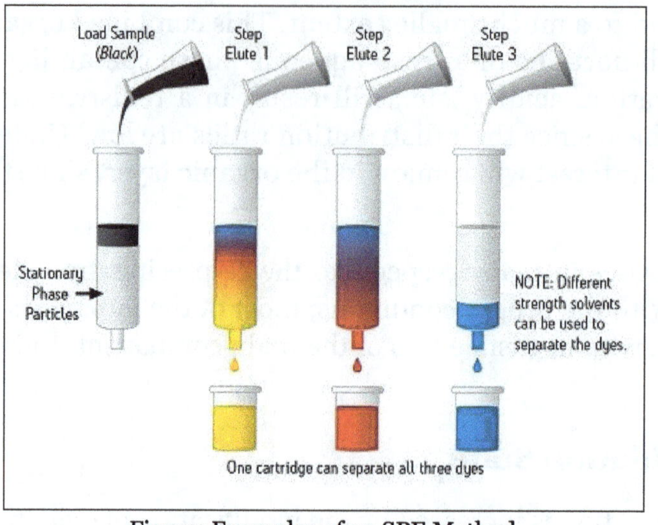

Figure: Examples of an SPE Method.

The chromatographic bed can be used to separate the different compounds in a sample, to make subsequent analytical testing more successful. For example, SPE is often used for the selective removal of interferences.

The technically correct name for this technology is "Liquid-Solid Phase Extraction," since the chromatographic particles are solid and the sample is in the liquid state. The same basic chromatographic principles of liquid chromatography that are used in HPLC are also used here, but in a different format and for a different reason. Here, chromatography is used to better prepare a sample before it is submitted for analytical testing.

In sample preparation, samples can come from a wide range of sources. They can be biological fluids such as plasma, saliva, or urine; environmental samples such as water, air, or soil; food products such as grains, meat, and seafood; pharmaceuticals; nutraceuticals; beverages; or industrial products. Even mosquito heads can be the sample. When a scientist needed to analyze neuropeptides extracted from the brains of mosquitoes, SPE was the sample preparation method of choice.

Four Major Benefits of SPE

There are many benefits to using SPE, but four major benefits deserve special attention.

Simplification of Complex Sample Matrix along with Compound Purification

One of the most difficult problems for an analytical chemist is when compounds of interest are contained in a complex sample matrix, such as mycotoxins in grains, antibiotic residues in shrimp, or drug metabolites in plasma, serum, or urine. The large number of interfering constituents or substances in the sample matrix along with the compounds of interest makes analysis extremely difficult.

The first problem to solve is the resulting complexity of the analysis itself due to the presence of so many entities which must be separated in order to identify and quantitate the compound[s] of interest.

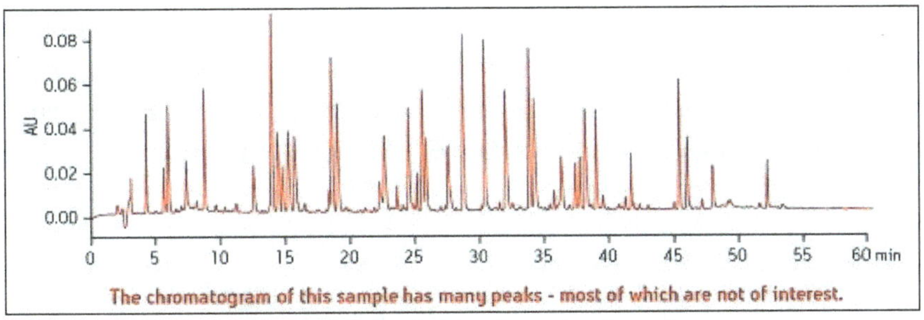

Figure: Example of a Complex Sample.

The robustness of the assay may not be adequate, because any slight change could impact the resolution of the separation of a critical pair of analytes.

Another consideration is that the presence of all the interferences in the original sample matrix can result in instrument downtime due to a buildup of contamination with each injection. If the interferences were removed as part of the sample preparation, then the compounds of interest could be analyzed with a simpler, more robust method. This can be seen in figure, which compares the original sample on top to the new SPE-prepared sample on the bottom.

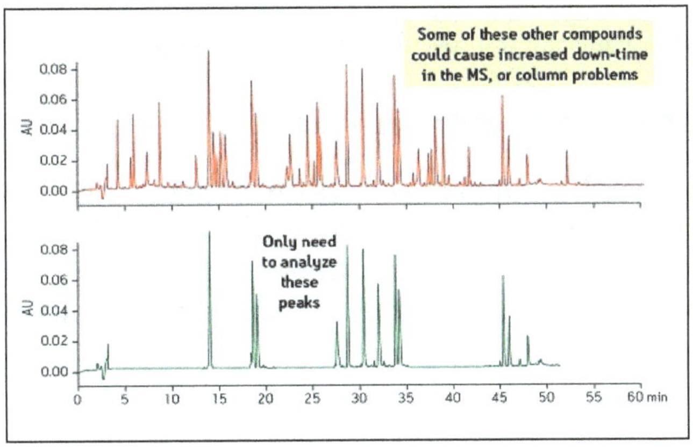

Figure: Comparison of Sample Matrix Complexities.

An additional benefit of simplifying the sample matrix is improved quantitation accuracy. The top blue trace for compound 1 in figure, initially appears to be acceptable. However, it really has some contamination from the sample matrix when compared to the blank sample matrix trace in red shown just beneath it. With a proper SPE protocol, the lower traces show the same compounds with no problems with interference, making quantitation much more accurate.

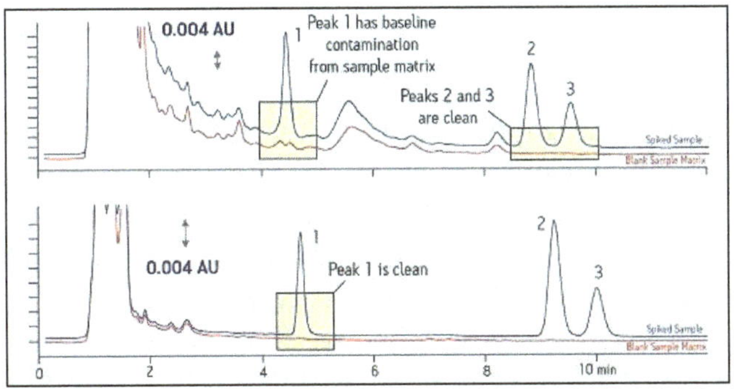

Figure: Improved Quantitation with Better Sample Preparation.

Another example is shown in figure. The upper trace shows significant interference from the sample matrix on both compounds 1 and 2. The lower trace shows much improved results [clean base line] due to proper sample preparation with SPE. Notice a much cleaner baseline improves the accuracy of the analytical results. Also, a much purer extract can be obtained if the sample requires isolation and purification of that compound.

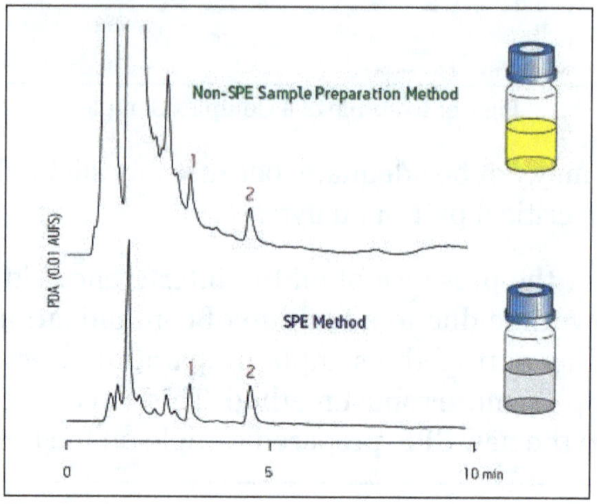

Figure: Significant Improvement in Baseline Using SPE Technology.

Reduce Ion Suppression or Enhancement in MS Applications

The second problem with complex sample matrices can be seen when we look at mass spectrometer output [LC/MS or LC/MS/MS]. For proper MS signal response [sensitivity], the compound ion must be allowed to form properly. In cases where the formation of the compound ion is suppressed by interferences in the sample matrix, the signal strength is greatly diminished.

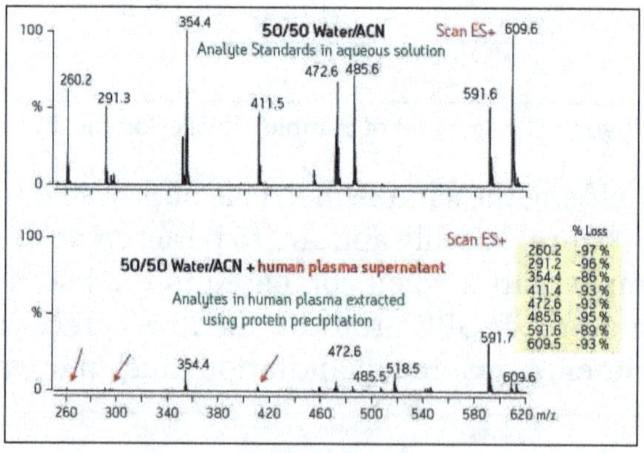

Figure: Example of Ion Suppression Due to Sample Matrix.

We can see this effect in figure. The upper output is the signal for our compounds of interest when injected in a saline solution. The lower trace shows significant reduction in response [> 90% suppression] of these same compounds when they were analyzed in human plasma. For the lower trace, only a common protein precipitation step was performed. This technique does not clean up the matrix interferences that cause the ion suppression, resulting in poor signal response.

Another good example of this suppression effect can be seen in figure. In the upper trace of the MS output, where the plasma sample was prepared with just a protein precipitation step, we can see that the terfenadine peak is suppressed by 80%. In the lower trace, where the same sample was prepared with a SPE method, we can see minimal ion suppression. Because the interferences from the sample matrix were removed, this allowed the compound ion to form properly, creating a better signal.

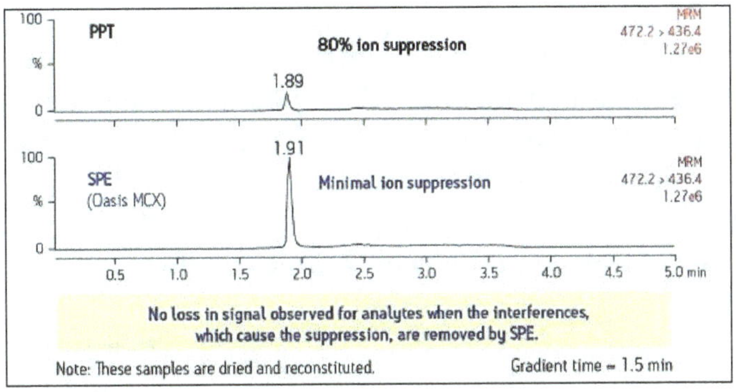

Figure: Reduced Ion Suppression with Proper SPE .

In some instances, interferences from the sample matrix can artificially increase the signal reported for a compound. This is called ion enhancement, resulting in an inaccurately high reported value. A proper SPE method will minimize this effect by cleaning away the interferences from the compound, resulting in a more accurate reported value.

Capability to Fractionate Sample Matrix to Analyze Compounds by Class

An analyst may be faced with a sample that contains many compounds, with a need to separate them by class so that further analysis can be carried out much more efficiently. For example, a soft drink beverage contains a wide range of compounds in its formulation. An SPE method could be developed to separate the different classes of compounds, for example by their polarity. The polar compounds could be collected, as a separated fraction, from the more non-polar compounds. These two fractions could then be separately analyzed in a much more efficient way because their compounds would be more similar.

An example of the power of fractionation by SPE is shown in figure below. Here, a complex sample of a dry powder [purple grape drink mix] is easily separated into four fractions: a fraction of just the polar compounds, a purified red compound, a purified blue compound, and a fraction containing all the remaining very non-polar compounds.

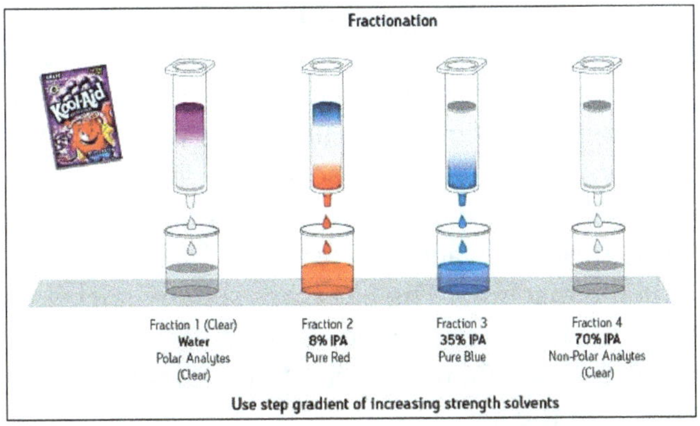

Figure: Sample Preparation by SPE.

Trace Concentration [Enrichment] of Very Low Level Compounds

Analysts today often need to report on compounds at far lower concentration levels than ever before, as little as parts per trillion [ppt] and even lower. Typically these levels are lower in the neat sample than the sensitivity capability of the analytical instruments.

A good example of this is the analysis for trace contaminants in environmental samples or metabolite development over time in biological fluids. The upper trace in figure shows the poor response of the original neat sample for the compound of interest. Using the same analytical conditions but with the sample prepared with SPE used in a trace concentration strategy, the lower trace shows a dramatic increase in signal strength for this compound. With this result, an accurate calculation of the original compound concentration in the neat sample can be made.

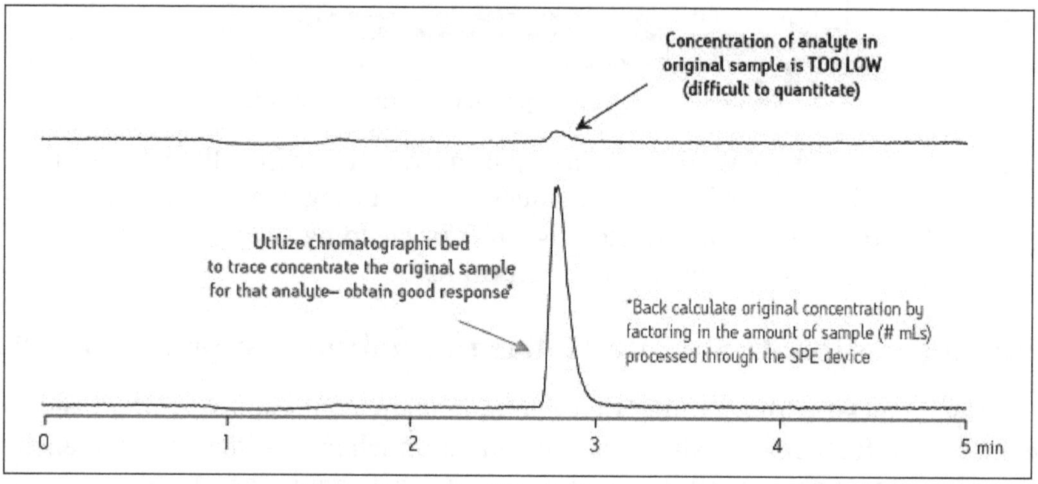

Figure: Example of Trace Concentration.

Without the retention capability of chromatographic packing materials in SPE, the ability to trace concentrate a specific compound[s] would be very difficult, if not impossible, with other sample preparation approaches.

Filtration

Filtration is one of the process of separation and it is defined as the action or process of filtering something. The act or process of filtering, especially the process of passing a liquid or gas, such as air, through a filter in order to remove solid particles. In liquid filtration Journals, such as that applied in waste material treatment, a liquid can be pulled through the filter by attraction, as in the examples given. On the opposite hand, some sort of applied pressure, or a pressure differential created by the existence of a vacuum, can force the liquid through the filter. Gas filtration is employed in a common appliance, the vacuum cleaner, which passes a stream of dust-filled air through a filtering bag inside the machine. The bag traps solid particles, while allowing clean air to pass back out into the room. This is essentially the same principle applied in air filters and even air conditioning and heating systems, which, in addition to regulating temperature, also remove dust, pollen, and other impurities from the air.

Filtration Techniques

Gravity Filtration

Gravity filtration uses a polyethylene or glass funnel with a stem and filter paper. Filter paper can have pore sizes ranging from small to large to permit slow to fast filtering. The paper is folded in half, then folded in quarters, and the tip of one corner is torn off to allow for a snug fit in the funnel cone. (If the paper has been pre-weighed, the torn corner piece must be saved to add to the post-filter weighing to avoid any errors.) The paper cone is fitted to the funnel so three thicknesses of the paper line one-half of the cone and one thickness lines the opposite half. Now place the funnel into a beaker and wet the filter paper completely with the dominate solvent or solvents in the mixture to be filtered.

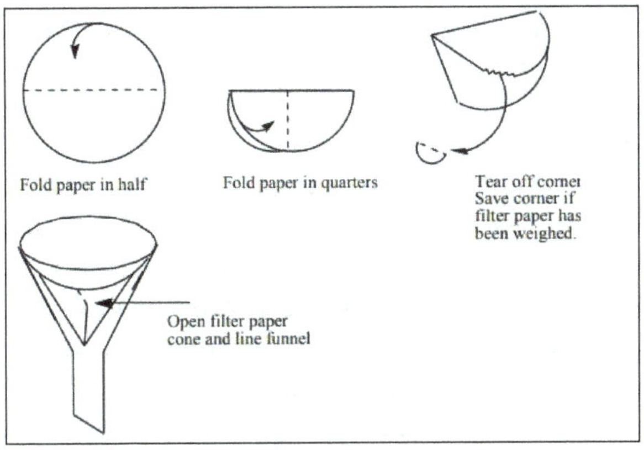

Figure: How to fold filter paper.

This step adheres the filter paper to the funnel walls preventing solid from escaping. Then, support the funnel with a clamp or ring (if necessary) and place a clean beaker beneath the funnel so the stem rests against the side of the beaker (this prevents splattering).

Before filtering, allow most of the solid in the mixture to settle. Now pour the supernatant liquid (the liquid standing over the solid in a mixture) through the filter first. This will allow the initial part of the filtration to proceed faster and may prevent clogging of the filter by the solid. To prevent splattering pour the liquid down a glass rod as shown in figure.

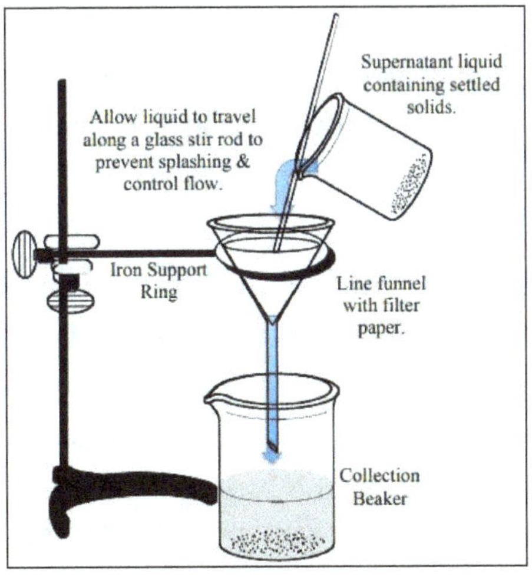

Figure: Gravity Filtration.

Scrape the solid onto the filter with a rubber policeman or spatula. Rinse the spatula, glass rod and beaker and pour the washings into the filter funnel. If the remaining solid residue is to be washed, rinse with three small portions (a few milliliters each) of an appropriate solvent. If the solid is to be saved, remove the filter paper carefully and place it on a watch glass to dry. Caution: Wet filter paper tears easily.

Sometimes the filtrate is tested to determine if a product or reactant has or has not passed through the filter. The test depends on what is being separated. For example: All barium ions $\left(Ba^{2+}\right)$ should have removed from solution by the formation of the precipitate $BaSO_4$. To check this a few drops of Na_2SO_4 solution can be added to a small portion of the filtrate. If no $BaSO_4$ precipitate forms, the filtration was successful. If a precipitate forms, additional precipitating (sulfate ion containing) reagents must be added to the filtrate and the resulting mixture must be filtered again.

Vacuum (or Suction) Filtration

Vacuum filtration uses a Buchner (pronounced "byook-ner") funnel and a water aspirator assembly. A Buchner funnel is a flat bottomed, porous, circular porcelain bowl with a short stem. The stem is fitted with a rubber stopper and inserted in the mouth of a side arm filter flask. Circular filter paper, the same diameter as the bowl, is placed on the flat bottom and wetted with the appropriate solvent to create a seal before starting the filtration.

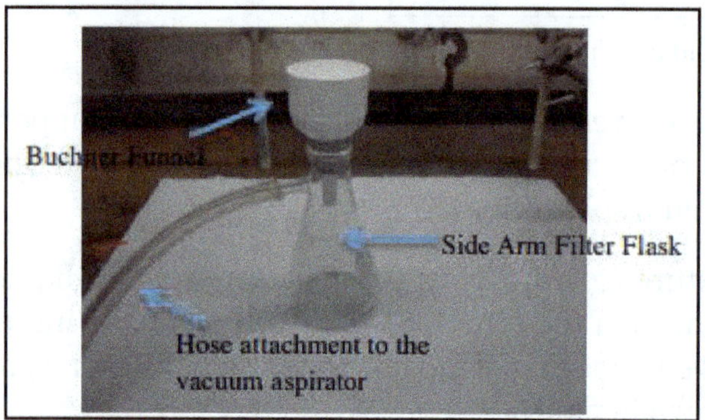

Figure: Buchner Funnel on a side arm filter flask.

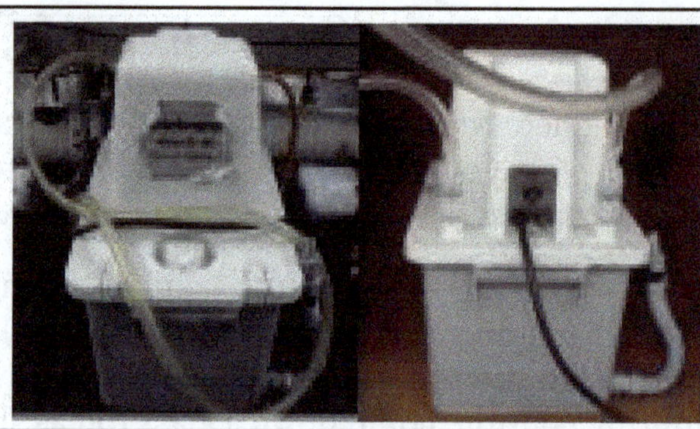

Figure: Vacuum Aspirator (front & rear views).

The hose in the above figure is attaches the side arm of the filter flask to a vacuum aspirator in the hood or at the lab bench. This vacuum aspirator creates the suction that pulls liquid through the filter and filter paper.

Figure shows the entire set-up consisting of a Buchner funnel, the side arm flask, and the vacuum aspirator. When the water aspirator is turned on the flow of water creates suction. The hose is part of the filtration system and should be heavy enough to prevent pinching or collapse under external atmospheric pressure.

To filter a sample, turn on the aspirator and carry out the filtration in the same manner described for gravity filtration. (Note: NEVER pry off the funnel if the system is under vacuum! Water can flash back into the collection flask or the filter paper can be damaged resulting in the loss of filtered solid.) Turn off the water aspirator before carefully removing the wet filter paper without tearing.

Reaction byproducts (either the solid or filtrate) should be placed into appropriate labeled containers in the hood.

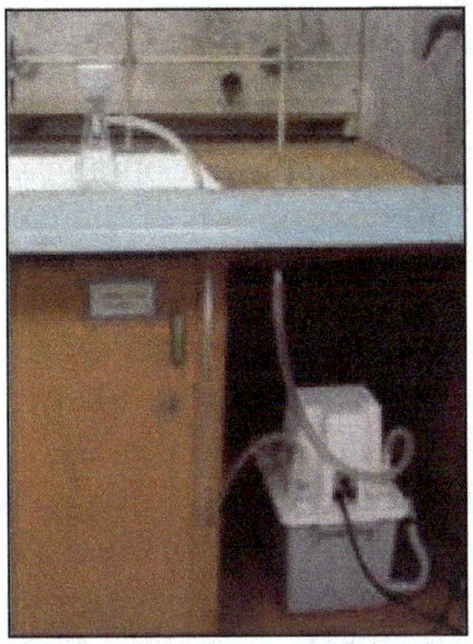

Figure: Vacuum Filtration Assembly.

Decantation

Decantation is a process to separate mixtures by removing a liquid layer that is free of a precipitate. The purpose may be to obtain a decant (liquid free from particulates) or to recover the precipitate. Decantation relies on gravity to pull precipitate out of solution, so there is always some loss of product, either from precipitate not fully falling out of solution or from leaving some liquid when separating it from the solid portion.

A piece of glassware called a decanter is used to perform decantation. There are several decanter designs. A simple version is a wine decanter, which has a wide body and a narrow neck. When wine is poured, solids stay in the base of the decanter. In the case of wine, the solid is usually potassium bitartrate crystals. For chemistry separations, a decanter may have a stopcock to drain away the precipitate or dense liquid or may have a partition to separate fractions.

Working of Decanting

There are two main decantation methods:

Separating Liquids and Solids

Decanting is done to separate particulates from a liquid by allowing the solids to settle to the bottom of the mixture and pouring off the particle-free part of the liquid.

For example, a mixture (possibly from a precipitation reaction) is allowed to stand so that gravity has time to pull the solid to the bottom of a container. The process is called sedimentation. Using gravity only works when the solid is less dense than the liquid. Clearwater may be obtained from mud by simply allowing time for the solids to separate from the water.

The separation may be enhanced using centrifugation. If a centrifuge is used, the solid may be compacted into a pellet, making it possible to pour off the decant with minimal loss of liquid or solid.

Separating Two or More Liquids

Another method is to allow two immiscible liquids to separate and the lighter liquid is poured or siphoned off. A common example is decantation of oil and vinegar. When a mixture of the two liquids is allowed to settle, the oil will float on top of the water so the two components may be separated. Kerosene and water can also be separating using decantation.

The two forms of decantation may be combined. This is especially useful if it's important to minimize loss of a solid precipitate. In this case, the original mixture may be allowed to settle or may be centrifuged to separate the decant and sediment. Rather than immediately drawing off the liquid, a second immiscible liquid may be added that is denser than the decant and that doesn't react with the sediment. When this mixture is allowed to settle, the decant will float on top of the other liquid and sediment. All of the decant can be removed with minimal loss of precipitate (except a tiny amount that remains floating in the mixture). In an ideal situation, the immiscible liquid that was added has a high enough vapor pressure that it evaporates, leaving all of the sediment.

Examples of Mixtures that can be Decanted

Some mixtures that can be decanted:

- Oil and water: Oil floats on top of water. Decanting the mixture allows the oil to be poured off the water.
- Gasoline or kerosene and water: This mixture is an example often cited as a safety hazard. Decanting a mixture containing flammable solvents can be dangerous as the flammable material evaporates and forms dangerous fumes.
- Dirt and water: Muddy water can be cleared up by decanting. The soil will sink to the bottom of the tube allowing the clear water to be poured off.
- Wine: Sediment from the fermentation process can produce an undesirable taste. Wine is decanted to separate the wine from these sediments.
- Cream and milk: Cream is separated from milk by decantation. Cream rises to the top of the milk mixture and is easily skimmed off.
- Blood and plasma: A centrifuge is necessary for this decantation. Plasma can be removed from blood by decantation.

Sublimation

Sublimation separates a mixture of solids, one of which sublimes. A few substances change directly, from a solid to vapor on heating without going through the liquid state. This change is called sublimation. On cooling, water vapor changes back to solid directly.

Iodine is a solid that sublimes. if a mixture of iodine and sand is heated in a beaker, the iodine changes from solid to vapor directly. This is a picture of sublimation process.

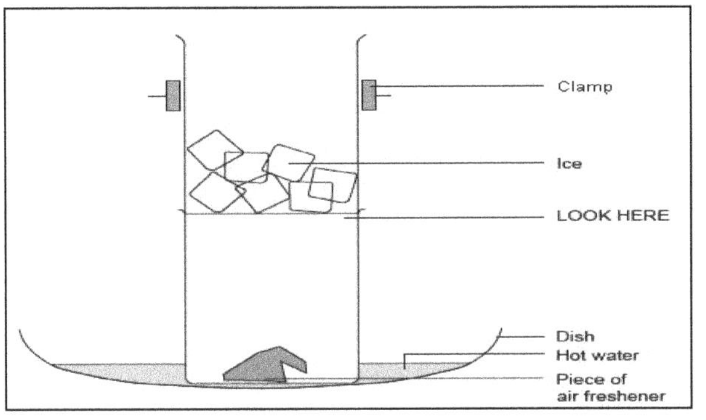

The vapor changes back to solid directly on a cold surface. The sand is not affected by the heat and remains in the beaker. Other substances that sublime include carbon dioxide, anhydrous iron (III) chloride, anhydrous aluminium chloride and some ammounium compounds. As only a few solids sublime, this method of separation is limited.

Examples of Sublimation

- Naphthalene: When used as insecticide and put between clothes, we can notice that it loses weight gradually. This loss of weight is due to sublimation. Naphthalene ball converts slowly to gas on standing. This gas is toxic to insects and hence they stay away from clothes.
- Water: Even water shows this behavior at triple point.
- Iodine: Iodine is a good example for sublimation.

Uses of Sublimation

- Drying: Sublimation helps to dry substance within getting moist. This process of sublimation is also beneficial in medicine. It is used in making powders from heat sensitive materials etc.
- Purification: If is a substance is in contaminated form, it can be separated into pure form by sublimation technique. Since the sublimation temperature of substance is different than the contaminant, when sublimation is done, the substance is separated leaving the contaminant behind.
- Aroma: Sublimation is the principle involved in air freshener solids used in room and toilets. The room fresheners are solids packed in a box. They spread the odor to the entire room by getting evaporated.

Evaporation

A puddle of water left undisturbed eventually disappears. The liquid molecules escape into the gas phase, becoming water vapor. Vaporization is the process in which a liquid is converted to a gas.

Evaporation is the conversion of a liquid to its vapor below the boiling temperature of the liquid. If the water is instead kept in a closed container, the water vapor molecules do not have a chance to escape into the surroundings and so the water level does not change. As some water molecules become vapor, an equal number of water vapor molecules condense back into the liquid state. Condensation is the change of state from a gas to a liquid.

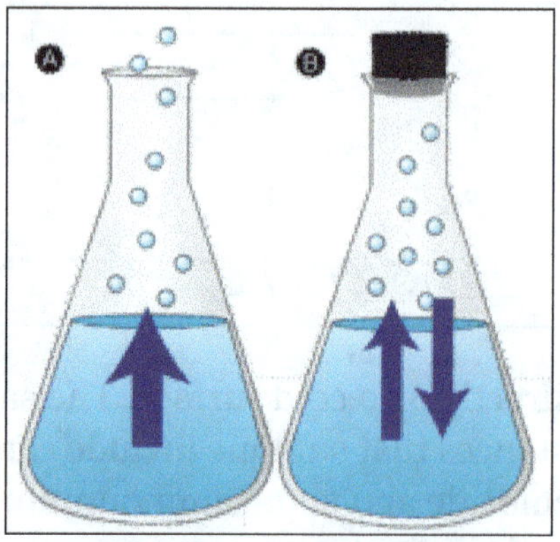

Figure: *Evaporation (A) and condensation (B).*

In order for a liquid molecule to escape into the gas state, the molecule must have enough kinetic energy to overcome the intermolecular attractive forces in the liquid. Recall that a given liquid sample will have molecules with a wide range of kinetic energies. Liquid molecules that have this certain threshold kinetic energy escape the surface and become vapor. As a result, the liquid molecules that remain now have lower kinetic energy. As evaporation occurs, the temperature of the remaining liquid decreases. You have observed the effects of evaporative cooling. On a hot day, the water molecules in your perspiration absorb body heat and evaporate from the surface of your skin. The evaporating process leaves the remaining perspiration cooler, which in turn absorbs more heat from your body.

A given liquid will evaporate more quickly when it is heated. This is because the heating process results in a greater fraction of the liquid's molecules having the necessary kinetic energy to escape the surface of the liquid. The figure below shows the kinetic energy distribution of liquid molecules at two temperatures. The numbers of molecules that have the required kinetic energy to evaporate are shown in the shaded area under the curve at the right. The higher temperature liquid (T_2) has more molecules that are capable of escaping into the vapor phase than the lower temperature liquid (T_1).

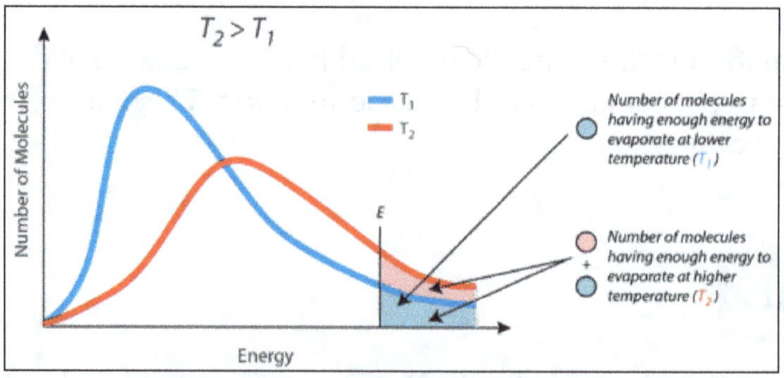

Figure: *Kinetic energy distribution curves for a liquid at two temperatures* T_1 and T_2. The shaded area is the molecules with enough kinetic energy to escape the liquid and become vapor.

Simple Distillation

Simple distillation is a procedure by which two liquids with different boiling points can be separated. Simple distillation (the procedure outlined below) can be used effectively to separate liquids that have at least fifty degrees difference in their boiling points. As the liquid being distilled is heated, the vapors that form will be richest in the component of the mixture that boils at the lowest temperature. Purified compounds will boil, and thus turn into vapors, over a relatively small temperature range (2 or 3 °C); by carefully watching the temperature in the distillation flask, it is possible to affect a reasonably good separation. As distillation progresses, the concentration of the lowest boiling component will steadily decrease. Eventually the temperature within the apparatus will begin to change; a pure compound is no longer being distilled. The temperature will continue to increase until the boiling point of the next-lowest-boiling compound is approached. When the temperature again stabilizes, another pure fraction of the distillate can be collected. This fraction of distillate will be primarily the compound that boils at the second lowest temperature. This process can be repeated until all the fractions of the original mixture have been separated.

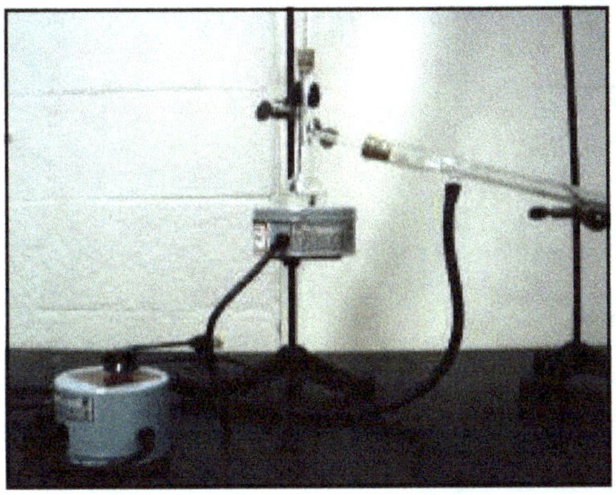

Figure: Distillation apparatus. A distillation flask with a thermometer is placed in a heating mantle and is connected to a condenser.

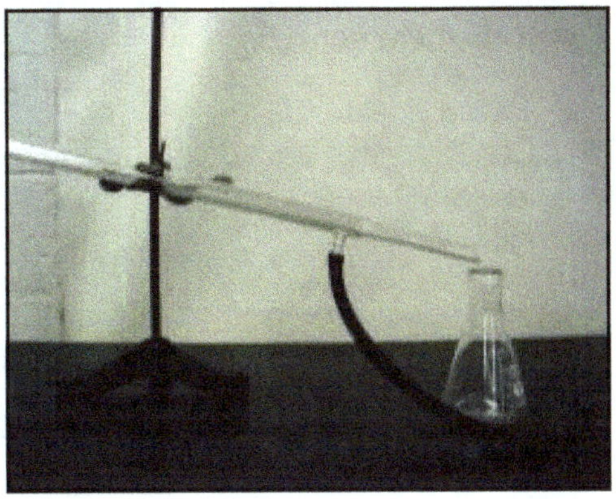

Figure: The tubes on the condenser are attached to a water source, with the water flowing in the low end and flowing out the high end of the condenser. The condensed vapor drips into the collection receiver.

Basic Procedure

1. Check the calibration of the thermometer that is to be used. This can be accomplished by placing the thermometer in an ice bath of distilled water. After the thermometer has been allowed to reach thermal equilibrium, place it in a beaker of boiling distilled water and again allow it to reach thermal equilibrium. If the temperatures measured deviate from the expected values by more than two degrees, obtain a new thermometer and check its calibration.

2. Fill the distillation flask. The flask should be no more than two thirds full because there needs to be sufficient clearance above the surface of the liquid so that when boiling commences the liquid is not propelled into the condenser, compromising the purity of the distillate. Boiling chips should be placed in the distillation flask for two reasons: they will prevent superheating of the liquid being distilled and they will cause a more controlled boil, eliminating the possibility that the liquid in the distillation flask will bump into the condenser.

Figure: The thermometer is inserted in the distillation flask through a hole in the cork stopper. The arm of the flask is inserted through a hole in the stopper of the condenser. Make sure these stoppers are airtight, or the vapor will escape.

3. Heat the distillation flask slowly until the liquid begins to boil. Vapors will begin to rise through the neck of the distillation flask. As the vapors pass through the condenser, they will condense and drip into the collection receiver. An appropriate rate of distillation is approximately 20 drops per minute. Distillation must occur slowly enough that all the vapors condense to liquid in the condenser. Many organic compounds are flammable and if vapors pass through the condenser without condensing, they may ignite as they come in contact with the heat source.

Figure: The distillation flask being heated in a heating mantle.

Figure: The collection receiver. The vapors condense and drip from the condenser into the flask.

4. As the distillate begins to drop from the condenser, the temperature observed on the thermometer should be changing steadily. When the temperature stabilizes, use a new receiver to collect all the drops that form over a two to three degree range of temperature. As the temperature begins to rise again, switch to a third collection container to collect the distillate that now is formed. This process should be repeated; using a new receiver any time the temperature stabilizes or begins changing, until all of the distillate has been collected in discrete fractions.

Note: All fractions of the distillate should be saved until it is shown that the desired compound has been effectively separated by distillation.

5. Remove the heat source from the distillation flask before all of the liquid is vaporized. If all of the liquid is distilled away, there is a danger that peroxides, which can ignite or explode, may be present in the residue left behind. Also, when all of the liquid has evaporated, the temperature of the glass of the filtration flask will rise very rapidly, possibly igniting whatever vapors may still be present in the distillation flask.
    - Never distill to dryness. The residue left in the distillation flask may contain peroxides, which could ignite or explode after all the liquid has distilled away.
    - Make sure that all joints are secured very tightly. If any vapor escapes at the connection points, it may come into direct contact with the heat source and ignite.
    - Never heat a closed system, the increasing pressure will cause the glass to explode. If the distillation flask has a tapered neck, the thermometer may be placed in such a way as to block to flow of vapors up the neck of the flask; in effect creating a closed system; make sure that if using a tapered neck flask, the thermometer is not resting in the lowest portion of the neck.

Simple distillation is effective only when separating a volatile liquid from a nonvolatile substance or when separating two liquids that differ in boiling point by 50 degrees or more. If the liquids comprising the mixture that is being distilled have boiling points that are closer than 50 degrees to one another, the distillate collected will be richer in the more volatile compound but not to the degree necessary for complete separation of the individual compounds.

The basic idea behind fractional distillation is the same as simple distillation only the process is repeated many times. If simple distillation was performed on a mixture of liquids with similar volatilities, the resulting distillate would be more concentrated in the more volatile compound than the original mixture but it would still contain a significant amount of the higher boiling compound. If the distillate of this simple distillation was distilled again, the resulting distillate would again be even more concentrated in the lower boiling compound, but still a portion of the distillate would be the higher boiling compound. If this process is repeated several times, a fairly pure distillate will eventually result. This, however, would take a very long time. In fractional distillation, the vapors formed from the boiling mixture rise into the fractionating column where they condense on the column's packing. This condensation is tantamount to a single run of simple distillation; the condensate is more concentrated in the lower boiling compound than the mixture in the distillation flask. As vapors continue to rise through the column, the liquid that has condensed will revaporize. Each time this occurs the resulting vapors are more and more concentrated in the more volatile substances. The length of the fractionating column and the material it is packed with impact the number of times the vapors will recondense before passing into the condenser; the number of times the column will support this is referred to as the number of theoretical plates of the column.

Since the procedures of simple distillation are so similar to those involved in fractional distillation, the apparatus that are used in the procedures are also very similar. The only difference between the equipment used in fractional distillation and that used in simple distillation is that with fractional distillation, a packed fractionating column is attached to the top of the distillation flask and beneath the condenser. This provides the surface area on which rising vapors condense, and subsequently revaporize.

The fractionating column is used to supply a temperature gradient over which the distillation can occur. In an ideal situation, the temperature in the distillation flask would be equal to the boiling point of the mixture of liquids and the temperature at the top of the fractionating column would be equal to the boiling point of the lower boiling compound; all of the lower boiling compound would be distilled away before any of the higher boiling compound. In reality, fractions of the distillate must be collected because as the distillation proceeds, the concentration of the higher boiling compound in the distillate being collected steadily increases. Fractions of the distillate, which are collected over a small temperature range, will be essentially purified; several fractions should be collected as the temperature changes and these portions of the distillate should be distilled again to amplify the purification that has already occurred.

Fractional Distillation

Fractional distillation is a special type of distillation designed to separate a mixture of two or more liquids that have different boiling points. The process involves heating the mixture and partial condensation of the vapors along a column (fractionating column), which is set up such that components with lower boiling points pass through the column and are collected earlier than components with higher boiling points. Generally, this method is used when the component parts boil at less than 25 °C from each other under a pressure of one atmosphere (atm).

The separation technology of fractional distillation is useful in both research and industrial settings. In industry, it is commonly used in petroleum refineries, chemical plants, and natural gas processing plants. Likewise, if a sample of air is liquefied, it can be separated into its components to produce liquid oxygen, liquid nitrogen, and argon. Also, chlorosilanes can be distilled to produce high-purity silicon for use as a semiconductor.

Fractional Distillation in a Laboratory

Apparatus

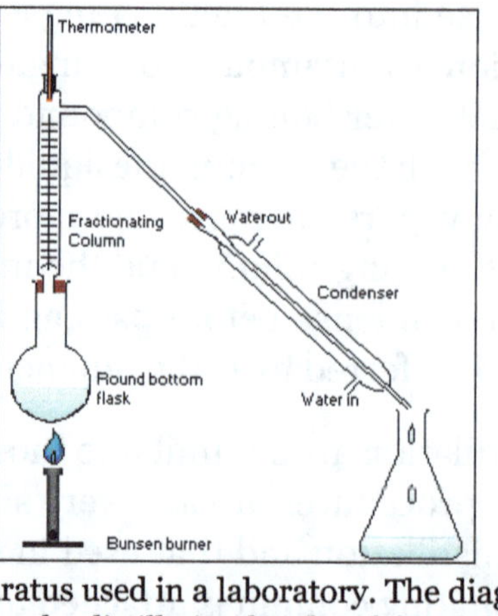

An example of a fractional distillation apparatus used in a laboratory. The diagram shows the use of a Liebig condenser and a conical flask as a receiving flask. Here, the distillation head and fractionating column are combined in one piece.

A laboratory setup for fractional distillation often involves use of the following pieces of apparatus, put together as shown in the image on the right:

- A heat source (such as a hot plate with a bath).
- A distilling flask (typically a round-bottom flask).
- A receiving flask (often a round-bottom flask or conical flask).
- A fractionating column (vigreux column).
- A distillation head.
- A thermometer and adapter if needed.
- A condenser (liebig condenser, graham condenser, or allihn condenser).
- A vacuum adapter (for distillations under reduced pressure).
- Boiling chips (also known as anti-bumping granules).
- Rubber bungs, unless laboratory glassware with ground glass joints is used, such as a quickfit apparatus.

Procedure

The apparatus is assembled as in the diagram (which represents a batch apparatus, as opposed to a continuous apparatus). The mixture is put into a round-bottom flask along with a few anti-bumping granules, and the fractionating column is fitted over the mouth of the flask. As the mixture boils, vapor rises up the column. The vapor condenses on the glass platforms, known as trays, inside the column, and runs back down into the liquid below. This is known as "refluxing" the distillate.

Only the most volatile fraction of the vapors remains in gaseous form all the way to the top of the column. This fraction passes into the condenser, which cools it down until it liquefies, and this liquid is collected in the receiving flask.

The efficiency in terms of the amount of heating and time required to get fractionation can be improved by insulating the outside of the column with an insulator such as wool, aluminum foil, or (preferably) a vacuum jacket. The hottest tray is at the bottom and the coolest is at the top. At steady state conditions, the vapor and liquid on each tray are at equilibrium. The fractionation is more thorough with the addition of more trays (up to certain practical limitations).

In laboratory distillation, any of several types of condensers may be used. The Liebig condenser is simply a straight tube within a water jacket, and is the simplest (and relatively least expensive) form of condenser. The Graham condenser is a spiral tube within a water jacket. The Allihn condenser has a series of large and small constrictions on the inside tube, each increasing the surface area upon which the vapor constituents may condense.

In alternate set-ups, a "cow" or "pig" adapter may be used, connected to three or four receiving flasks. By turning the "cow" or "pig," the distillates can be channeled into the appropriate receiver. A Perkin triangle may also be used to collect distillation fractions, without requiring a "cow" or "pig" adapter. A Perkin triangle is most often used when the distillates are air-sensitive, or when the fractions distill and are collected under reduced pressure, but it can be used for simple as well as fractional distillations.

Vacuum distillation systems operate at reduced pressure, thereby lowering the boiling point of the materials.

Example

Consider the distillation of a mixture of water and ethanol. Ethanol boils at 78.5 °C, and water boils at 100 °C. On that basis, one should be able to separate the two components by fractional distillation. However, a mixture of 96 percent ethanol and four percent water boils at 78.2 °C, being more volatile than pure ethanol. Such a mixture is called an azeotrope. When the mixture is gently heated, the azeotrope (being the most volatile component) concentrates to a greater degree in the vapor and separates from the rest of the liquid first. Thus, fractional distillation of a mixture of water and ethanol produces 96 percent ethanol. Once all the ethanol has boiled out of the mixture, the thermometer shows a sharp rise in temperature.

Industrial Distillation

Typical industrial fractional distillation columns.

Fractional distillation is the most common form of separation technology used in petroleum refineries, petrochemical and chemical plants, and natural gas processing plants. For example, this technique is used in oil refineries to separate crude oil into useful substances (or fractions) consisting of different hydrocarbons with differing boiling points. The crude oil fractions with higher boiling points consist of larger molecules (with more carbon atoms and higher molecular weights), are darker in color, are more viscous, and are more difficult to ignite and to burn.

Most often, new feed is continuously added to the distillation column, and products are continuously removed. Unless the process is disturbed due to changes in feed, heat, ambient temperature, or condensing, the amount of feed being added and the amount of product being removed are normally equal. This is known as continuous, steady-state fractional distillation.

Industrial distillation is typically performed in large, vertical cylindrical columns known as "distillation or fractionation towers" or "distillation columns." Their diameters range from about 65 centimeters to six meters, and their heights range from about six meters to 60 meters or more. The distillation towers have liquid outlets at intervals up the column, allowing for the withdrawal of different *fractions* or products with different boiling points or boiling ranges. The "lightest" products (those with the lowest boiling point) exit from the top of the columns and the "heaviest" products (those with the highest boiling point) exit from the bottom of the column.

Fractional distillation is also used for the separation of (liquefied) air into its components, producing liquid oxygen, liquid nitrogen, and high purity argon. Distillation of chlorosilanes enables the production of high-purity silicon for use as a semiconductor.

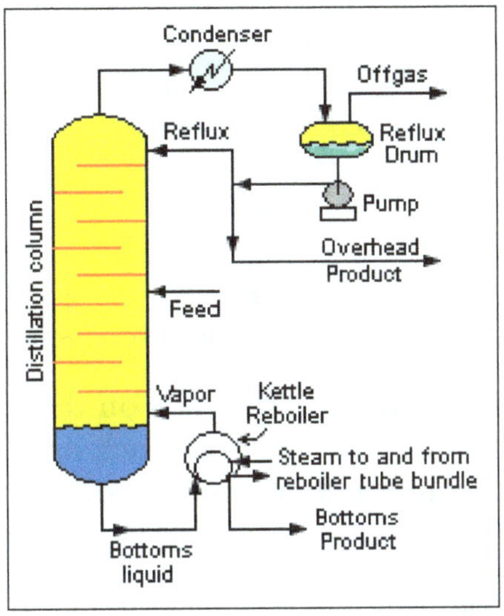

Diagram of a typical, large-scale industrial distillation tower.

Large-scale industrial towers use reflux to achieve a more complete separation of products. Reflux refers to the portion of the condensed overhead liquid product from a fractionation tower that is returned to the upper part of the tower as shown in the schematic diagram on the right. Inside the tower, the reflux liquid flowing downward provides the cooling needed to condense the vapors flowing upward, thereby increasing the effectiveness of the distillation tower. The more reflux is provided for a given number of theoretical plates, the better the tower's ability to separate lower boiling materials from higher boiling materials. Alternatively, the more reflux provided for a given desired separation, the fewer theoretical plates are required.

In industrial uses, sometimes a packing material is used instead of trays within the column, especially when operating under reduced pressures. This packing material can either be random dumped packing (one to three inches wide) such as Raschig rings or structured sheet metal. Liquids tend to wet the surface of the packing and the vapors pass across this wetted surface, where mass transfer takes place. Unlike conventional tray distillation, in which every tray represents a separate point of vapor liquid equilibrium, the vapor liquid equilibrium curve in a packed column is continuous. However, when modeling packed columns, it is useful to compute a number of "theoretical plates" to denote the separation efficiency of the packed column with respect to more traditional trays. Differently shaped packings have different surface areas and void space between packings. Both of these factors affect packing performance.

Design of Industrial Distillation Columns

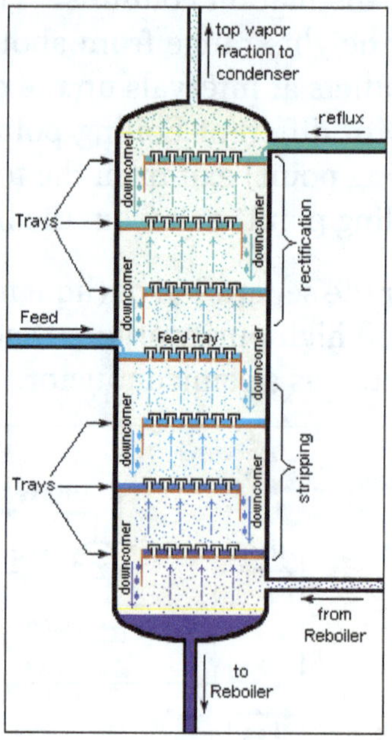

Chemical engineering schematic of typical bubble-cap trays in a distillation tower.

Design and operation of a distillation column depends on the feed and desired products. Given a simple, binary component feed, analytical methods such as the McCabe-Thiele method or the Fenske equation can be used. For a multi-component feed, simulation models are used both for design and operation.

Moreover, the efficiencies of the vapor-liquid contact devices (referred to as plates or trays) used in distillation columns, as seen in Figure two, are typically lower than that of a theoretical 100 percent efficient equilibrium stage. Hence, a distillation column needs more plates than the number of theoretical vapor-liquid equilibrium stages.

Centrifugation

Centrifugation separates on the basis of the particle size and density difference between the liquid and solid phases. Sedimentation of material in a centrifugal field may be described by,

$$V = \frac{d^2(\rho_s - \rho_I)\omega^2 r F_s}{18\eta\theta}$$

where v is the rate of sedimentation, d is the particle diameter, ρ_s is the particle density, ρ_I is the solution density, w is the angular velocity in radians s^{-1}, r is the radius of rotation, h is the dynamic viscosity, F_s is a correction factor for particle interaction during hindered settling and q is a shape factor (=1 for spherical particles). F_s depends on the volume fraction of the solids present; approximately equalling 1, 0.5, 0.1 and 0.05 for 1%, 3%, 12% and 20% solids volume fraction respectively. Only material which reaches a surface during the flow through continuous centrifuges will be removed from the centrifuge feedstock, the efficiency depending on the residence time within the centrifuge and the distance necessary for sedimentation (D). This residence time will equal the volumetric throughput (f) divided by the volume of the centrifuge (V). The maximum throughput of a centrifuge for efficient use is given by,

$$\phi = \frac{d^2(\rho_s - \rho_I)\omega^2 rVF_s}{18\eta\theta D}$$

The efficiency of the process is seen to depend on the solids volume fraction, the effective clarifying surface (V/D) and the acceleration factor (w^2r/g, where g is the gravitational constant, 981 cm s^{-2}; a rotor of radius 25 cm spinning at 1 rev s^{-1} has an acceleration factor of approximately 1 G). Low acceleration factors of about 1 500 g may be used for harvesting cells whereas much higher acceleration factors are needed to collect enzyme efficiently. The product of these factors (w^2rV/gD) is called the sigma factor (S) and is used to compare centrifuges and to assist scale-up.

Laboratory centrifuges using tubes in swing-out or angle head rotors have high angular velocity (w) and radius of rotation (r) but small capacity (V) and substantial sedimentation distance (D). This type of design cannot be scaled-up safely, primarily because the mechanical stress on the centrifuge head increases with the square of the radius, which must increase with increasing capacity.

For large-scale use, continuous centrifuges of various types are employed. These allow the continuous addition of feedstock, the continuous removal of supernatant and the discontinuous, semi-continuous or continuous removal of solids. Where discontinuous or semicontinuous removal of precipitate occurs, the precipitate is flushed out by automatic discharge systems which cause its dilution with water or medium and may be a problem if the precipitate is required for further treatment. Centrifugation is the generally preferred method for the collection of enzyme-containing solids as it does not present a great hazard to most enzymes so long as foam production, with consequent enzymic inactivation, is minimised.

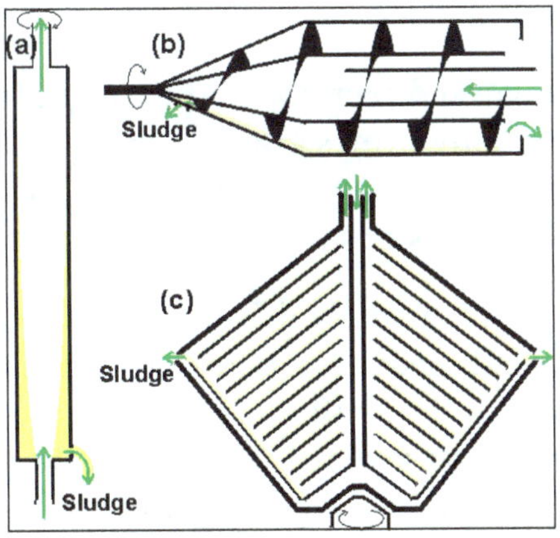

In figure basic designs of industrial centrifuges, showing the flow of material within the bowls. Motor drives, cooling jackets and sludge collection vessels are not shown. (a) Tubular bowl centrifuge. This is generally operated vertically, the tubular rotor providing a long flow path enabling clarification. The sludge collects and must be removed. (b) Continuous scroll centrifuge. This is operated horizontally. The helical screw scrolls the solids along the bowl surface and out of the liquid; the sludge being dewatered before discharge. The clarified liquor overflows over an adjustable weir at the other end of the bowl. The screw conveyer rotates at a slightly different speed to the bowl. (c) Continuous multichamber disc-stack centrifuge. The bowl contains a number of parallel discs providing a large clarifying surface with a small sedimentation distance. The sludge is removed through a valve.

Centrifuges

Centrifuge is any device that applies a sustained centrifugal force—that is, a force due to rotation. Effectively, the centrifuge substitutes a similar, stronger, force for that of gravity. Every centrifuge contains a spinning vessel; there are many configurations, depending on use. A perforated rotating drum in a laundry that throws off excess water from clothes, for example, is a type of centrifuge. A similar type is used in industry to separate fluids from solid matter after crushing.

As enunciated by Sir Isaac Newton in his first law of motion, a freely moving body (such as a ball) tends to travel in a straight line, and, if directed along a curved path by some restraining force (such as would result were a hand-held string tied to it), it will exert a force against the directing or restraining force in its continual effort to fly off onto a straight tangential course. It is a familiar observation that an object revolving in a circle exerts a force away from the center of rotation. This force, which is the outward pull of the ball on its string, is the centrifugal force. Also, there is general appreciation of the fact that the amount of this force can be increased by increasing either the angular velocity of rotation, the mass of the object, or the radius of the circle through which the object moves. Perhaps not so generally appreciated is the fact that, whereas the centrifugal force is directly proportional to the radius and to the mass, it is proportional to the square of the angular velocity. For example, doubling the mass of the rotating object will increase the centrifugal force by a factor of 2, but doubling the number of revolutions per minute (rpm) will increase the centrifugal force by a factor of 4 (equals 2 times 2); similarly, increasing the speed by a factor of 10 will increase the force by a factor of 100 (equals 10 times 10). Centrifugal force is expressed by the basic relation $F = mv^2 / R = 4\pi^2 mn^2 R$; F is the centrifugal force, m the mass, R the radius, v the speed, and n the number of revolutions per second.

The centrifugal force is often compared directly with the weight (pull of gravity) of the object, and the amount of force is stated as so many "times gravity" or so many "g." Through the use of special research apparatus, forces greater than 5,000,000 times gravity have been produced by spinning small metal rotors of about pea size at speeds exceeding 1,000,000 revolutions per minute.

The rotating element of a centrifuge is usually driven about a fixed axis by an electric motor, or by an air turbine in some high-speed machines, and is known as a rotor, bowl, or drum. For the minimizing of vibration and strain on the shaft and bearings, it is essential that a loaded rotor be well balanced—i.e., that its total mass be so distributed about the axis of rotation that the resultant of all the elemental forces is zero. If the bearings are suited to high speeds and if ample power is available to overcome the frictional resistance of the bearings, the only limitation to the speed of a well-balanced rotor is the strength against rupture of the material from which it is made.

For example, a rotor with a 15-cm (6-inch) diameter used in certain biological studies and designed especially for high speeds has a limiting speed for routine operation of about 60,000 revolutions per minute. In a rotor of given design, the maximum angular velocity obtainable before rupture is to a close approximation inversely proportional to the rotor's diameter. Thus, a small rotor having only one-half the diameter of a larger one can be as safely rotated at twice the angular velocity and with the production at the periphery of twice the centrifugal force.

The widest use of centrifuges is for the concentration and purification of materials in suspension or dissolved in fluids. Suspended particles denser than the suspending liquid tend to migrate toward the periphery, while those less dense move toward the center. The rapidity with which the migration proceeds is dependent on the intensity of the centrifugal field, the difference between the density of the particle and that of the suspending liquid, the viscosity of the liquid, the size and shape of the particle, and to some extent the concentration of the particles and the degree to which they are electrically charged. The net motivating force exerted on the particle is the difference between the centrifugal field acting on it and the opposing buoyancy of the liquid. All other things being equal for two particles, one with a diameter 10 times that of the other will require only $^1/_{100}$ as much average centrifugal field to move a given distance in a given time as the smaller.

From the foregoing discussion, it is clear that a practically complete separation of the suspending medium and the suspended particles can be produced if the centrifugation is allowed to continue until all particles have collected against the outer wall of the spinning vessel or centrifuge. It should also be noted that a partial separation of two groups of suspended particles of different size can be effected by allowing centrifugation to continue only long enough for all of the larger particles to be completely packed into the sediment, since then many of the small particles will still be suspended in the fluid. If separation of the larger as well as the smaller particles is desired, the surface fluid can be drawn off and the sediment re-suspended in some suitable liquid and subsequently centrifuged again to effect further separation.

Centrifuges may be classified in three general categories depending on whether the spinning centrifuge bowl that contains the material to be separated has a solid wall, a perforated wall, or some combination of the two. Also, they may be characterized according to whether the material is treated in a continuous flow process, a batch process, or a combination of the above processes.

Bottle Centrifuges

A bottle centrifuge is a batch-type separator that is primarily used for research, testing, or control. The separation takes place in test tube or "bottle-type" containers, which are symmetrically mounted on a vertical shaft. The shaft of a bottle centrifuge is usually driven by an electric motor, gas turbine, or a hand-driven gear train located above or below the rotor. In most cases, the bottles are supported by high-strength metal containers so that their axis is perpendicular to the axis of rotation. The sedimentation occurs in a radial direction, and in some bottle centrifuges the test tubes or bottles are inclined at an angle of about 37° to the axis of rotation in order to reduce the distance that the material must settle.

Bottle centrifuges are standard equipment for most biological, chemical, or medical laboratories. They are used to separate solid materials in suspension or to clarify liquids when precipitation will not take place in a reasonable time in the gravitational field g. In most commercial bottle centrifuges the centrifugal field may be varied from a few g up to tens of thousands. Commercial uses of the bottle centrifuge include tests for the butterfat content of milk, determination of the sediment in crude mineral and vegetable oils, and clinical tests of various kinds.

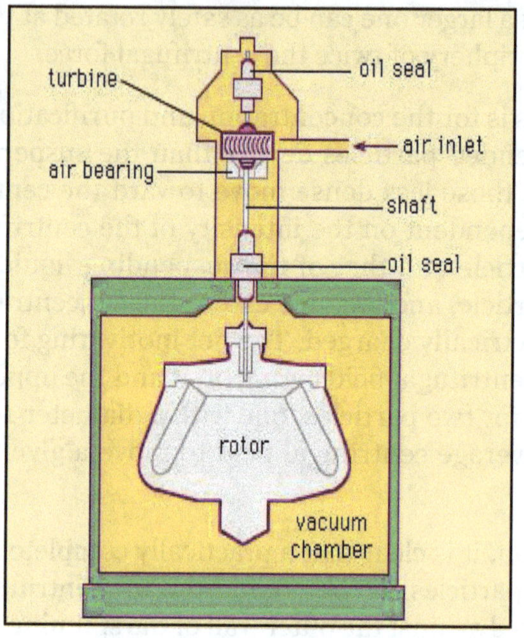

Tubular Centrifuges

The tubular centrifuge is used primarily for the continuous separation of liquids from liquids or of very fine particles from liquids, although in some cases it is employed as a batch-type centrifuge. In general, it is used when higher centrifugal fields are required for separation. The rotating bowl of a tubular centrifuge consists of a long hollow tube (length many times its diameter) as shown schematically in figure. For continuous separation the feed or material to be centrifuged enters at one end near the axis and is removed in two streams containing the separated material. In many cases the separation is not complete, and the separated fractions must be passed through the machine several times. Many different designs for the internal structure of the tube are employed, but, in general, radial vanes are used to bring the feed material up to speed and to slow down the separated streams before they are discharged. The centrifuge is driven by a high-speed motor or an air or steam turbine. The sedimentation takes place as the fluid flows from one end of the tube to the other. When the heavy material consists of very fine particles or molecules and the concentration is very low, the solid material is usually allowed to deposit on the wall. In this case the machine is operated as a batch centrifuge.

The tubular centrifuge is finding an increasing number of applications because of the high centrifugal fields that may be used (10^5g near the periphery in some cases). A few typical uses are as follows:

1. The purification of vaccines (un-centrifuged vaccines contain a large amount of nonessential and harmful material).
2. Purification of lubricating and industrial oils.
3. Clarification and purification of food products such as essential oils, extracts, and fruit juices.
4. Separation of immiscible liquids that cannot be separated by gravity.

Disk-Type Centrifuges

The disk-type centrifuge consists of a vertical stack of thin disks in the shape of cones. The sedimentation takes place in the radial direction in the space between adjacentcones. This greatly reduces the settling distance and hence increases the rate at which the material is separated. The angle of the cones is designed so that upon reaching the inside surface of the cone the heavier material slides down along its surface in a manner that is similar to that of the 37° fixed-angle bottle centrifuge.

The disk-type centrifuge usually operates continuously. The material to be processed enters in one stream and is separated into two purified streams. These centrifuges are used primarily for the separation of liquids in which the solid or immiscible components occur in relatively low concentrations. The familiar cream separator, widely used in the dairy industry and on farms for separating cream from milk, is a typical example of this type of centrifuge. They also are used for the purification of fuel oil, the reclamation of used motor oil, and the removal of soap stock in the refining of vegetable oils.

Basket Centrifuges

Basket centrifuges are often called centrifugal filters or clarifiers. They have a perforated wall and cylindrical tubular rotor. In many cases the outer wall of a basket centrifuge consists of a fine mesh screen or a series of screens with the finer mesh screens supported by the heavier coarse screen, which in turn is supported by the bowl. The liquid passes through the screen, and the particles too large to pass through the screen are deposited. The basket centrifuge is employed in the manufacture of cane sugar, in the home and in laundries for the rapid drying of clothes, and in the washing and drying of many kinds of crystals and fibrous materials, etc.

Vacuum-Type Centrifuges

In the centrifuges described above, the rotor spins in air or some other gas at atmospheric pressure. The gaseous friction on a spinning rotor increases at a relatively high rate so that the power required to drive the rotor also increases rapidly. As a result, the temperature of the rotor rises drastically, sometimes exceeding the boiling point of water. As the rotor surface near the periphery moves faster than near the axis, a thermal gradient or variation in temperature through the rotor wall is established along the radius with the periphery at a higher temperature than the axis. These small radial temperature gradients produce convection within the centrifuge, and these convection currents can cause remixing and disturb sedimentation.

The heat buildup and convection problems caused within a centrifuge by air resistance can be avoided by spinning the rotor within an evacuated chamber. The elimination of air resistance also makes possible the attainment of high rotational speeds with relatively little expenditure of energy. Many vacuum-type centrifuges are ultracentrifuges; i.e., they operate at speeds of more than about 20,000 revolutions per minute. figure shows a schematic diagram of an early vacuum-type ultracentrifuge. The centrifuge rotor located inside the vacuum chamber is connected to the air-supported, air-driven turbine by a vertical, small-diameter, flexible steel shaft.

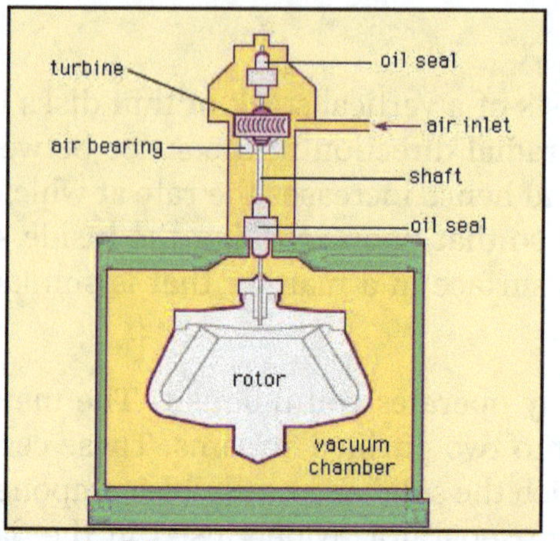

Figure: Air-driven, vacuum-type ultracentrifuge.

The rotor of a typical vacuum-type ultracentrifuge is 18 cm (7 inches) in diameter and carries 300 ml (10 ounces) of liquid in a centrifugal field of more than 300,000*g*. Practically all substances of importance in medicine and biology and all other substances with molecular weights of 50 daltons (one dalton is 1.66×10^{-24} grams) or more are easily purified in this type of bottle centrifuge. The rotor of a vacuum-type ultracentrifuge can be replaced by one with sector-shaped cells and transparent windows so that the progress of the sedimentation can be optically measured and photographed. This method was first used by T. Svedberg and J.B. Nichols in 1923 and was widely applied thereafter to determine the sedimentation rates and sizes of many submicroscopic particles, particularly protein molecules and viruses.

The vacuum-type centrifuge may be used for the determination of the molecular weights of practically all substances in solution. In modern commercial vacuum-type centrifuges the air drive and support have been replaced by the more efficient and convenient electric motor drive, and the entire machine has been redesigned and made almost automatic in its operation. The present commercial vacuum-type ultracentrifuge has become an indispensable tool in laboratories where it is necessary to purify substances of importance in biochemistry, biophysics, biology, medicine, and the pharmaceutical industry.

The ultracentrifuge can be used in two principal ways for determining the molecular weights of various proteins. The first consists in carrying out the sedimentation in a centrifugal field high enough to produce a relatively sharp sedimentation boundary–i.e., the boundary between the sedimenting molecules and the pure solvent. The rate at which this boundary moves out along the radius toward the periphery is then measured and the value of the molecular weight is calculated. This is called the rate of sedimentation method. The second method consists in centrifuging the material until equilibrium is established in the centrifuge cell–i.e., until the rate at which the material settles out is balanced by back diffusion. If the concentration in the cell is then determined at various radial distances, the value of the molecular weight can be calculated.

Vacuum-type tubular centrifuges are used to purify many biological materials that cannot easily be separated in other ways. They have been employed both as continuous-flow and as density-gradient centrifuges. The density-gradient centrifuge consists in setting up a radial density variation or gradient in the tubular centrifuge with slowly sedimenting nonreactive smaller molecules such as sucrose or calcium chloride. If, then, the density of the substance to be purified falls within the range of the artificial density gradient, it will collect in a thin cylindrical surface at a definite radius. If more than one substance is in the solution, each of the substances will collect at a radius determined by its particle density.

Another important use of the vacuum-type centrifuge is gas separation. When a gas is subjected to a centrifugal field, a radial pressure gradient is immediately established. Consequently, a mixture of any two gases with different molecular weights may be separated in a centrifuge with the lighter gas being concentrated on the axis. In 1919, after it was pointed out that it should be possible to separate the isotopes of an element by centrifuging, a number of attempts were made to obtain separation but were all unsuccessful, probably due to convection and remixing in the centrifuge. In 1937 the isotopes of chlorine were separated with a vacuum-type ultracentrifuge. An evaporative centrifuge method was used in which the material to be separated is admitted to the rotor and condensed on the periphery with the rotor stationary. The rotor is then driven to operating speed and the lighter material pumped out through the hollow shaft while the heavier material remains in the centrifuge to be collected later. The centrifuge used in gas separation should be spun as rapidly as possible and should be as long as possible. The centrifuge method is suited to the separation of the heavier isotopes as well as the lighter ones, because it depends on the differences in the masses rather than on their absolute values.

Since the mid-1940s the technique of gaseous centrifuging has been further developed and extended. Workers in Germany and in the Netherlands have had considerable success with the method. A remarkably simple vacuum-type gas centrifuge that is especially adapted to uraniumisotope separation has been devised. During the 1970s a centrifuge plant was constructed in Europe for the purpose of commercially producing reactor-grade uranium-235 for use in nuclear power plants.

Magnetic Separation

The basic principle behind magnetic separations is remarkably simple and remains unchanged from these early examples. It relies on the simple fact that materials with differing magnetic moments experience different forces in the presence of magnetic field gradients; thus, an externally applied field can handpick out of physically similar mixtures those components with distinctive magnetic characteristics. The use of this principle is straightforward in mixtures where a magnetic component is known to exist; an intrinsically magnetic material can be separated using electrically powered electromagnets or strong permanent magnets. The process is generally binary and results in a magnetically rich retentate (usually a solid) and the residual non-magnetic solvent.

The 1950s were a time of great expansion for the field of magnetic separations as the introduction of high gradient magnetic separation (HGMS) systems permitted faster and more general magnetic separation processes. HGMS works through the application of large static fields ($\sim 1\ T$) to columns containing ferrous matrices such as steel wool; these irregular surfaces give rise to magnetic gradients as high as 10^4 T/m which generate forces large enough to capture even weakly magnetic particles in a flow stream. This new technology meant that magnetic separations could be applied more universally to separation problems where particulate matter was not strongly magnetic. The use of HGMS in water treatment, for example, provided a way to clarify water with high sediment loads, provided that the sediment has some weakly magnetic character. A larger market was found in the steel industry where HGMS separators became critical components for purifying high grade, low iron content steel.

More recently, separations using external magnetic fields have become commonplace in biotechnology where they are used for both protein purification as well as flow cytometry. In these contexts, separators apply relatively small field gradients (500 T/m) to fixed volumes of solution with the aim of removing a valuable biological components. Such an approach to separations is more as a `batch process', and it works well for biomedical research laboratories where solution volumes are small and separation speed is not a major issue. However, the small scale and time consuming nature of biomagnetic separators precludes their application in biological manufacturing processes. In addition, the separators themselves are standalone configurations of permanent magnets which require little upkeep or even power. This separator simplicity is offset by the need for magnetic beads that are coated so as to bind to particular biological components. As a result, the technique is not that general and conventional biomedical researchers can only use magnetic separations if a commercial magnetic bead source is available for their targets. Commercial sources for magnetic beads have grown substantially in the past decade, however, and a wide variety of proteins, cells and other biomacromolecules may be selectively removed using these methods.

Magnetic Separations: Their Unique Position among Separation Technologies

The problem of separating materials, whether they are specialized chemicals, high purity steel or valuable pharmaceuticals, from product streams is a near universal one for any manufacturing process. When speed is not an issue and the materials of interest are solids or flocculated products, sedimentation or centrifugation is routinely employed. For faster processing, filtration is a well-established method for removing waste or concentrating product. Given these standard methods it is not always obvious that magnetic separations could or should be applied to a given problem. Certainly, if part of a mixture is intrinsically magnetic then magnetic removal is often the best solution. In these cases, magnetic separations generally offer higher throughput with greater specificity than equivalent centrifugation or filtration methods.

However, even in the absence of intrinsically magnetic components the use of designer magnetic beads—targeted to the product of interest—can make a magnetic separation feasible for virtually any system. Such processes offer very different kinds of tradeoffs in speed and selectivity as opposed to the more conventional approaches. A consideration of the unique advantages of moving materials with external fields, as opposed to other conventional alternatives, is detailed in table. For the purposes of this analysis we considered both the slower `batch' model for magnetic separations as well as the faster flow separators based on the high gradient magnetic columns. We also limit our comments to separation processes for materials in liquids, a case which captures the majority of current application in this area.

An important feature of a magnetic separation is that the product stream meets with virtually no flow resistance as it moves through a separator; this is in stark contrast to filters which use a solid phase as the basis for the separation process. Filtration is particularly challenging when submicron particles or biomolecules are the target products; this requires filter pore sizes well under a micron (ultrafiltration) or even a submicron (nanofiltration). The flow resistance that this solid phase introduces to the system is significant, and industrial membrane filters often must operate with hundreds of pounds of solution pressure (many times the flow of a fire hydrant) to maintain flow rates of several milliliters per minute. Such units require not just energy-intensive pumping stations, but also expensive and high performance fittings and valves. Filters also can become fouled with a wide variety of contaminants requiring backwashing procedures and ultimately replacement.

Conventional magnetic separators face few of these problems. In high-gradient flow systems, the columns are loosely packed with steel wools that offer virtually no resistance to flow. Reasonable operating pressures and conventional pumps are suitable for these systems, though the electromagnets that are needed for very high gradient separations do require significant energy. For separations in which speed is not an issue, then the batch magnetic separators are interesting options. Because they usually generate the fields with permanent magnets, they can operate without any external power; with the appropriate beads, such systems can remove a greater variety of substances than those accessed through sedimentation. This makes them particularly attractive for small-volume separations in low infrastructure settings.

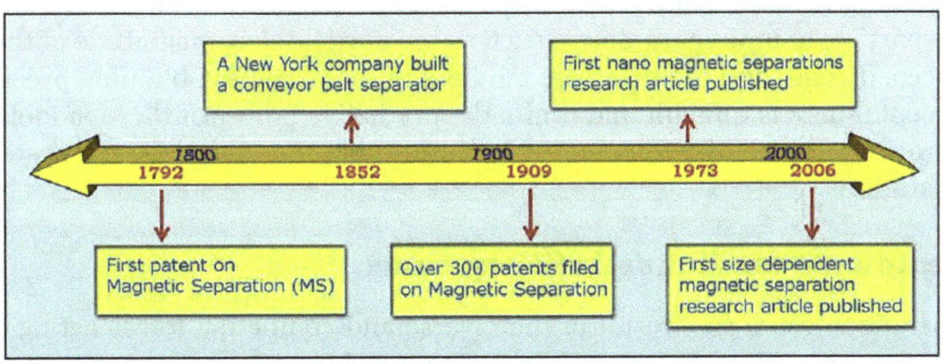

Figure: The evolution of magnetic separation technology.

Table: Industrial applications of magnetic separations

Application area	Objective
Chemical and related industries	Tramp metal removal from machinery to avoid wear and malfunction
Food industries	
Coal processing plants	
Metals production and recycling industries	
Raw materials processing plants, such as cement, glass, semiconductors	Removal of iron and derivatives contamination
Mineral post processing industries	Separation and enrichment of magnetic ores (magnetite, hematite, ilminite, etc.)

While magnetic separations are in principle excellent choices for many problems, their widespread application has been limited by the complexity required in separator design and bead technology. In the case of flow separations, cumbersome and expensive electro- magnets are used to create external fields in excess of a Tesla; the requirement for high gradients also limits column diameters and thus reduces throughput. In the batch separations favored in biotechnology, permanent magnets in a multipole configuration provide the relatively low gradients needed and their set-up and use is easy and immediate. The trade-off here is that these low fields produce only very small magnetic forces on particles. Since the magnetic force is proportional to the particle volume, the small field gradients are generally offset by the use of larger magnetic beads. This lowers the available surface area for biomolecular recognition and adsorption, and has precluded application of this technology to biological separations on the larger scale. Our group and others are working on ways to better design nano scale magnetic beads so as to overcome these and other limitations. This materials design process must be guided by a more quantitative and formal description of the process of magnetic separations.

Modeling Magnetic Separations

The magnetic separation process is amenable to simple modeling and such treatments date back to the 1950s; most analyses assume that the materials being removed act independently and that macroscopic models for fluid dynamics are appropriate. These constraints make it straightforward to describe the many forces acting on particles in a flow, such as the magnetic force, Brownian motion, gravitation, and friction. What results is a universal equation that relates the net force on a particle to input parameters such as the particle size, magnitude of the field gradient, frictional coefficient, and the magnetic moment of the particle. Obtaining precise values for these experimental data is difficult and limits the predictive power of these models. Still, these treatments provide an essential foundation for improving the materials and systems used for magnetic separations.

Simple Treatments for Magnetic Separations

Magnetic separation occurs as a result of the diverse and competing forces acting on a particle during flow in the presence of an external field. These include a hydrodynamic drag force due to the flow velocity; a magnetic force due to the gradient of the applied magnetic field; gravitational forces for large and dense particles; and finally a diffusive force which, arises from the intrinsic Brownian motion of particles. While the first three are deterministic forces that control a particle's trajectory, the latter is a randomizing force that must be exceeded if an ensemble of particles is to experience net movement during some time interval. Put in other terms, if the energy available to move a particle does not exceed its thermal energy due to Brownian motion then it is not possible to set up a concentration gradient.

Perhaps the most important factor to include in models of magnetic separations is the role of inter-particle forces such as Helmhotz double layer interaction, dipole–dipole interaction and Van Der Waals attraction. Thus, the magnetic, dipole–dipole interaction and Van Der Waals forces aid the process of separation, whereas, diffusion, double layer interaction and drag force act against the separation. Also, the particles are found to reversibly aggregate in presence of high field gradients, which would increase the net magnetic moment to a magnitude much stronger than the competing forces.

Magnetic Separations use in Column Formats: Examples

Magnetic separators used on the industrial scale are overwhelmingly high gradient systems which function as a column in continuous flow operations.

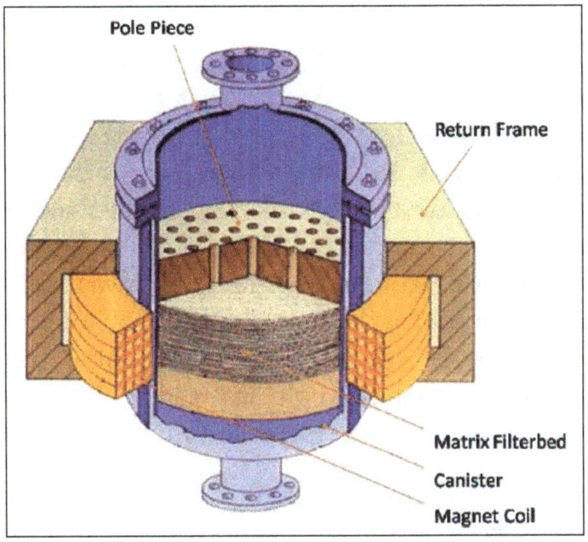

Figure: Metso® high gradient magnetic separators (HGMS) are designed to recover weakly magnetic material from non-magnetic matter and can be used for many applications including the processing of clays, iron ores, rare earths and industrial minerals. In addition to the strongly magnetic minerals of Fe, Co, and Ni, a vast number of weakly magnetic minerals, which are not normally treatable by ordinary magnetic separators, may be processed by high gradient magnetic separators. Metso HGMS separators are able to remove even weakly paramagnetic materials (from Metso minerals).

The speed and efficiency of these systems have been exploited for decades and the early stages of the industry were marked by notable optimism: "Virtually every process in the chemical engineering industry is a potential application (for HGMS). Many previously unthinkable processes will now become practical, and many previously practical ones will become unthinkable. It has already happened in the kaolin industry, and is beginning to hap- pen elsewhere." While these systems are not as ubiquitous now as their inventors anticipated thirty years ago, their applications have expanded notably.

Kaolin (Clay) Decolorization

Kaolin (a.k.a. china clay) is a clay mixture primarily consisting of kaolinite ($Al_2O_2SiO_2.2H_2O$ or $Al_2Si_2O_5(OH)_4$) mineral. It is named from the Kaoling Hills of the city of Chingte chen where fine Chinese porcelains were produced; its resistance to acids and alkalis alike was greatly prized. Today, however, it's mostly used in the paper manufacturing industry where it plays dual role, as a filler between the pulp fibers and as a surface coating for a white glossy finish.

Natural kaolin has color as mined due to the iron containing micas, tourmaline, pyrite, anatase and rutile present in the material. To remove these impurities, kaolin can be magnetically cleaned with a continuous high gradient magnetic separator to produce highly white material suitable for paper or porcelain. Figure shows kaolin mineral before and after the decolorization process. Because of the resistance of kaolin impurities to other chemical cleaning methods, HGMS handles 75% of the world production of white porcelain and paper (Oder, 1976). A typical plant would have an HGMS with a filter diameter of about 2 m and capacity up to 20 t/h.

On average, generating 1 t of steel requires 151 t of water for cooling and cleaning purposes; the resulting wastewater is filled with many magnetic particulates and other iron-containing impurities. Those particles, especially when present in gas and hot water streams, cause significant problems in processing and must be removed. Conventional methods for cleaning steel mill waste and process waters include sedimentation, flocculation followed by sedimentation, and fixed bed filtration. Such approaches require either large areas for settling tanks and clarifiers or expensive and short- lived filter systems. Magnetic separation has emerged as an ideal solution for this industry and it has offered great time, space and cost savings. In a sample treatment at Kawasaki Steel Corporation of Japan, a 3 kOe field strength, 2.1 m diameter magnetic filter removes 80% of contaminants from the cooling wastewater of vacuum degassing process. Similar use in treatment of wastewater can also be found in power plants (both conventional and nuclear). For these cases, HGMS is used to remove ferromagnetic or paramagnetic particulates which extends the lifetime of cooling systems.

Magnetic separations can also be used to treat pollution. Fly ash from coal power plants is 18% iron oxide. Magnetic filtration has been applied to capture 15% of waste fly ash, thus providing a means for recycling. Estimates show that this can replace some of the magnetite used in industry. Figure shows an example of a ball mill separator used in these operations.

Enrichment of Ores–mineral Beneficiation

The treatment of ores with magnetic separation is carried out primarily to enrich iron-containing ores. Conventional chemical and settling methods are not suited for this purpose given the similar density and reactivity of transition metal minerals. The magnetic nature of iron species, however, is unique and thus a natural target for magnetic separations. Among the iron ores taconite is most often subjected to magnetic treatments. From a taconite ore (33% iron) Kelland (1973) was able to recover iron at 95% on a 5 cm/s flow rate. Today, Metso Minerals, Inc. (formerly Sala International AB) offers magnetic separators that can separate iron from ores with nearly 100% efficiency (depending on the particulate sizes, magnetic field and flow rate). Figure shows a successful continuous HGMS separator used for these purposes.

Magnetic separation of pyrite (FeS2) from coal for desulfurization is also a common process. The weakly magnetic nature of pyrite, however, requires that the raw ore be pre-treated thermally to convert the pyrite (FeS_2, Ms = 0.3 emu/g) to more strongly magnetic pyrrhotite (Fe_7S_8, Ms = 22 emu/g). Up to 91% removal of sulfur from coal can be achieved by microwave heating followed by a magnetic separation. Figure shows an industrial scale drum separator used in large scale applications for powders such as coal.

Food Industry

Strict food quality standards require the food products to be contaminant free, where mainly rare earth elements (REEs) constitute the majority. The food industry, therefore, has found magnetic separations to be an ideal method to remove REEs from food ingredients. Similar to the ore beneficiation or desulfurization, the target substances are weakly magnetic and require the high gradients of a magnetic field to be removed in a continuous food production line. Bunting Magnetics Co. offers magnetic metal separators and metal detectors for the quality of food and extended service life of the processing equipment, especially for cheese processing, chocolate plants, pet food processing, flour mills, spice plants, vegetable processing.

Figure: Kaolin decolorizes to white after magnetic separation.

Figure: A ball mill separator. Separation of ball mill grinding ball segments from the discharge.

Removal of both ferrous and nonferrous tramp metals is achieved by their line of food safety products for the food processing industry. Figure shows case studies from Greenwood Magnetics Ltd., another company that produces magnets and assembly systems for cleaning REEs off of the food production lines.

Water Treatment and Metal Removal

With the new, lowered maximum permissible concentration for arsenic in drinking water (10 μg/l), effective since January 2006, techniques for better arsenic remediation without much desorption have gained more importance. Currently, coprecipitation, adsorption in fixed-bed filters, membrane filtration, anion exchange, electrocoagulation, and reverse osmosis are of methods of interest, however, cost efficiency and waste quantity requires further development that would aid in resolving the problem. Arsenic adsorption and desorption are heavily influenced by ad- sorbent particle size. Nanoscale magnetite (Fe_3O_4, 12 nm) can remove 200 times better than its commercial counterparts, which allows a significant cut in waste, instead of 1.4 kg of bulk iron oxide to remove arsenic (500 μg/l) from 50 l of solution, 15 g of nano magnetite can be used.

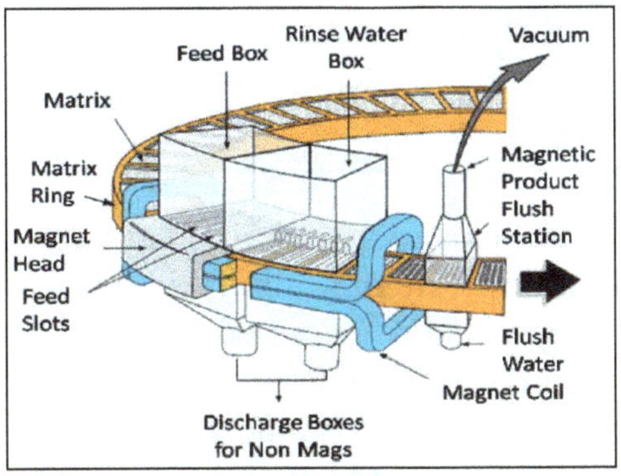

Figure: Continuous high gradient magnetic separation for many low susceptibility minerals that are associated with other minerals or have extra Fe in the crystals, and are hence often possible to separate.

Apart from surface area increment, an obvious gain while going down to nanoscale, available open sites with the proper chemistry (free Fe on the surface) can be accounted for this unexpected result. Size dependent magnetic properties bring controllability and along with mobility, a critical nanoscale advantage, provides unique applicability, if put in a system, in especially household uses where electricity is not readily available.

As early as 1970s Delatour (1973) and Delatour and Kolm (1975) treated water samples from the Charles River (Fe_3O_4 seeding, 5 ppm Al^{3+}) with a high flow velocity HGMS (V_o = 136 mm/s, H_o = 1 T) and reduced coliform bacteria from 2.2×105/l to 350/l, turbidity by 75%, color by 95%, and suspended solids by 78%. Later, Bitton and Mitchell removed 95% of the viruses from water by magnetic filtration following a 10 min of contact period with magnetite (added to be 250 ppm). The following years, Boliden Kemi AB reduced phosphorus of water supplies at least 87%. Also known as Sirofloc process, micron-sized magnetite is also used to remove color, turbidity, iron, and aluminum from water sources as an alternative to metal-ion coagulation. Recently, Denizli, and coworkers magnetically modified yeast cells for facile capture of mercury with fast biosorption rates (within 60 min) and efficiency (76.2 mg/g for Hg^{2+}).

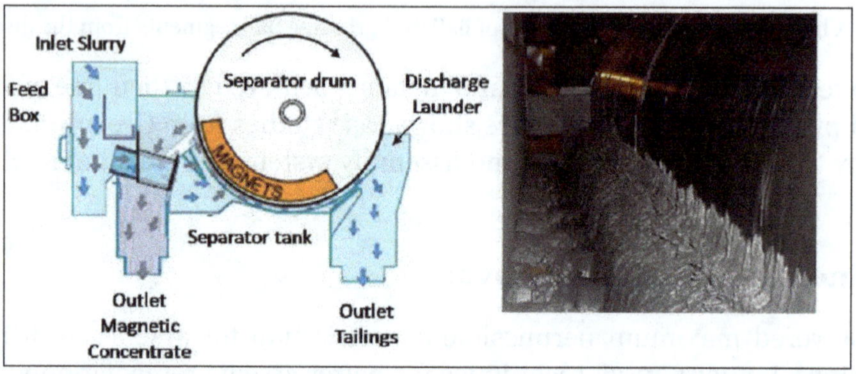

Figure: An industrial scale drum separator used in large scale applications.

(Left) A single row easy-clean grid box which contains high-density rare earth easy-clean magnetic tubes filters loose tea with a flow rate of 5 tph. (Middle) A water-jacketed pipeline magnet was manufactured to suit a 4 in pipeline pressure resistant up to 10 bar, nine high intensity rare earth magnets (11 500 G) filter liquid chocolate flowing at 300 l/min. The pressurized heated water-jacket maintains the temperature of the chocolate. (Right) The Bullet magnet that is used by a flour producer in which flour flows upwards through the 5 in pipe and any ferrous contamination is removed by the high intensity rare earth bullet magnet (8500 G min).

Table: Sources of contaminants in a steel production process.

Source of contaminant	Contaminants
Coke production	Non-magnetic particles, organics, and oils
Iron manufacturing	Magnetic particles and organics
Steel production	Magnetic particles
Cold finishing	Magnetic particles, oils, and acids
Hot formation	Magnetic particles and oils

Table: Arsenic removal by nano-sized magnetite.

Particle size (nm)	As (V) or As (III)	Residual as concentration (μg/l)	% Removal
12	As (III)	3.9	99.2
20	As (III)	45.3	90.9
300	As (III)	375.7	24.9
12	As (V)	7.8	98.4
20	As (V)	17.3	96.5
300	As (V)	354.1	29.2

References

- Chromatography: explainthatstuff.com, Retrieved 04, February 2020
- Applications-of-chromatography: chromatography.conferenceseries.com, Retrieved 10, March 2020
- Liquid-chromatography: chemicool.com, Retrieved 18, May 2020
- Definition-of-decantation-604990: thoughtco.com, Retrieved 11, April 2020
- Sublimation-decanting-and-centrifuging: theartofseparationtechniques.blogspot.com, Retrieved 21, July 2020
- What-is-sublimation: studyread.com, Retrieved 25, August 2020
- Fractional-distillation: newworldencyclopedia.org, Retrieved 16, January 2020

Permissions

All chapters in this book are published with permission under the Creative Commons Attribution Share Alike License or equivalent. Every chapter published in this book has been scrutinized by our experts. Their significance has been extensively debated. The topics covered herein carry significant information for a comprehensive understanding. They may even be implemented as practical applications or may be referred to as a beginning point for further studies.

We would like to thank the editorial team for lending their expertise to make the book truly unique. They have played a crucial role in the development of this book. Without their invaluable contributions this book wouldn't have been possible. They have made vital efforts to compile up to date information on the varied aspects of this subject to make this book a valuable addition to the collection of many professionals and students.

This book was conceptualized with the vision of imparting up-to-date and integrated information in this field. To ensure the same, a matchless editorial board was set up. Every individual on the board went through rigorous rounds of assessment to prove their worth. After which they invested a large part of their time researching and compiling the most relevant data for our readers.

The editorial board has been involved in producing this book since its inception. They have spent rigorous hours researching and exploring the diverse topics which have resulted in the successful publishing of this book. They have passed on their knowledge of decades through this book. To expedite this challenging task, the publisher supported the team at every step. A small team of assistant editors was also appointed to further simplify the editing procedure and attain best results for the readers.

Apart from the editorial board, the designing team has also invested a significant amount of their time in understanding the subject and creating the most relevant covers. They scrutinized every image to scout for the most suitable representation of the subject and create an appropriate cover for the book.

The publishing team has been an ardent support to the editorial, designing and production team. Their endless efforts to recruit the best for this project, has resulted in the accomplishment of this book. They are a veteran in the field of academics and their pool of knowledge is as vast as their experience in printing. Their expertise and guidance has proved useful at every step. Their uncompromising quality standards have made this book an exceptional effort. Their encouragement from time to time has been an inspiration for everyone.

The publisher and the editorial board hope that this book will prove to be a valuable piece of knowledge for students, practitioners and scholars across the globe.

Index

www.ingramcontent.com/pod-product-compliance
Lightning Source LLC
LaVergne TN
LVHW080726210926
848485LV00003B/524